Ergebnisse der Anatomie und Entwicklungsgeschichte
Reviews of Anatomy, Embryology and Cell Biology
Revues d'anatomie et de morphologie expérimentale

Herausgegeben von

A. Brodal, Oslo · C. Elze, München · W. Hild, Galveston · R. Ortmann, Köln
G. Töndury, Zürich · E. Wolff, Paris

Schriftleitung
G. Töndury, Zürich

Band 39 · Heft 2

Rolf Nyberg-Hansen

Functional Organization
of Descending Supraspinal Fibre Systems
to the Spinal Cord

Anatomical Observations
and Physiological Correlations

With 39 Figures

Springer-Verlag Berlin Heidelberg GmbH 1966

Rolf Nyberg-Hansen, Anatomisk Institutt, Universitetet i Oslo,
Oslo 1/Norge, Karl Johans gate 47 (Domus Media)

ISBN 978-3-540-03494-0   ISBN 978-3-662-30435-8 (eBook)
DOI 10.1007/978-3-662-30435-8

Contents

I. Introduction

Recent advances in the neurophysiology of the spinal cord, due largely to the use of microelectrodes, have increased the demand for a detailed knowledge of its minute anatomy, including the exact sites and mode of termination of the various contingents of afferent fibres to the spinal grey matter, among them the descending supraspinal fibre systems. Anatomical data of this kind are indispensable for functional interpretations and for the analysis of the structural and functional organization of the spinal cord.

The observation of REXED (1952, 1954) that the grey matter of the feline spinal cord may be subdivided on a cytoarchitectonic basis into ten different laminae, presumably representing, at least in part, functionally different regions, should serve as a stimulus to attempt more precise analysis of the intrinsic organization of the spinal cord. Furthermore REXED's laminae provide a common basis of reference of the sites of termination of afferent fibre systems to the spinal grey matter and the localization of single units recorded from in neurophysiological experiments, and thus promise useful correlations between anatomical and physiological observations and their functional interpretations.

The terminal regions within the spinal grey matter of the various descending supraspinal fibre systems have been studied in different animals and in man by previous authors using the Marchi method, the RASDOLSKY method and various silver impregnation techniques. Since, however, the Marchi method gives information on myelinated fibres only, and since these fibres loose their myelin sheaths at their terminal ramifications, this method does not permit detailed analysis of the sites of termination of degenerating fibres. In the reports of most of the students employing silver techniques many details which are of utmost interest for functional interpretations have regrettably not been considered. Most previous authors indicate the terminal regions of the various descending supraspinal fibre systems in rather general terms, usually speaking of the "ventral horn", "intermediate zone", "basal parts of the dorsal horn" and the "dorsal horn" as being the recipient regions. It cannot, however, always be gathered from the text and illustrations how the author deliminates these regions of the spinal grey matter. Regions regarded as belonging to the "ventral horn" by one author are by other authors referred to as parts of the "intermediate zone", and so forth.

In previous studies, furthermore, little or no attention has been devoted to the synaptic relationships. Thus most authors do not distinguish between the ventral motor horn cells and the other types of nerve cells located in the ventral horn.

In view of the state of affairs outlined above a detailed anatomical mapping of the sites of termination in the spinal cord of the various descending supraspinal fibre systems in the cat was undertaken, with special reference to the laminar organization described by REXED (1952, 1954). Furthermore, by employing the silver impregnation methods of NAUTA (1957) and GLEES (1946) information on the synaptic relationships has been achieved, especially when segments of the spinal cord are cut in the transverse as well as the horizontal and sagittal planes.

Below the principal findings of experimental investigations on seven descending supraspinal fibre systems (NYBERG-HANSEN and BRODAL 1963, 1964; NYBERG-HANSEN and MASCITTI 1964; NYBERG-HANSEN 1964a, b; 1965a; 1966) will be reported following a few remarks on the material and methods employed. The main purpose of this review will be to attempt a general discussion with special reference to correlations between anatomical and neurophysiological observations which are of relevance for the understanding of the structural and functional organization of the spinal cord. Emphasis will be put on synthesis and integration rather than on detailed documentation.

Reference to the literature and detailed discussions of the author's findings in relation to those of previous authors' will be limited to a minimum, since complete references and discussions are found in the original papers on which the present account is based.

II. Material and methods

Altogether 90 adult cats have been used as experimental animals. Under Nembutal anesthesia and with sterile precautions lesions have been made in the cerebral cortex by dissection with a knife and sucker, and in various nuclei and regions of the brain stem by means of a Horsley-Clarke stereotaxic instrument. The electrodes used were insulated with lacquer but for the tip. All needle tracks were carefully controlled microscopically, and usually turned out to be discrete. Most animals were treated postoperatively with penicillin (300.000 I.U.).

The cats were killed 5 to 11 days after the operations by an overdose of Nembutal and intravital perfusion with 10 per cent formalin. The brain and the spinal cord were then fixed in 10 per cent formalin for an adequate period of time.

The part of the brain containing the lesion was cut in transverse (frontal) sections, and stained with thionine for identification of the lesions.

Selected segments from various levels of the spinal cord were cut serially on the freezing microtome, and impregnated according to the silver techniques of NAUTA (1957), NAUTA-GYGAX (1954) and GLEES (1946), as reported in more detail in the original publications on which the present account is based. An evaluation of the criteria employed in the identification of degenerating fibres and terminals is likewise found in these papers. "Terminal" degeneration thus denotes the degenerative changes occurring in the terminal boutons and the fairly fine axonal ramifications leading up to them, while "preterminal" degeneration indicates degenerating fibres, usually somewhat coarser than the terminal ones.

The various cytoarchitectonic laminae of the spinal grey matter described by REXED (1952, 1954) can be identified in transverse as well as in longitudinal sections, and the borders between the various laminae identified in 100 μ thick Nissl sections can be transferred to drawings of silver impregnated sections. Furthermore, the different laminae present certain characteristics in silver impregnated sections which facilitate the identification of the various laminae as described in the original publications.

III. Results

a) The corticospinal fibre system

From previous experimental studies it appears likely that almost the entire cerebral cortex contributes fibres to the corticospinal fibre system, although from a quantitative point of view the primary sensorimotor area is the most important (see NYBERG-HANSEN and RINVIK 1963, for a recent review). Accordingly, the lesions in our experimental animals were made in various regions of this cortical area. A few cases from the original study (NYBERG-HANSEN and BRODAL 1963) will be described to show the essential findings.

Fig. 1 shows the findings in cat B.St.L. 214 (killed after 6 days) in which the lesion includes the whole primary sensorimotor area, and in addition destroys

Cat B.St.L. 214

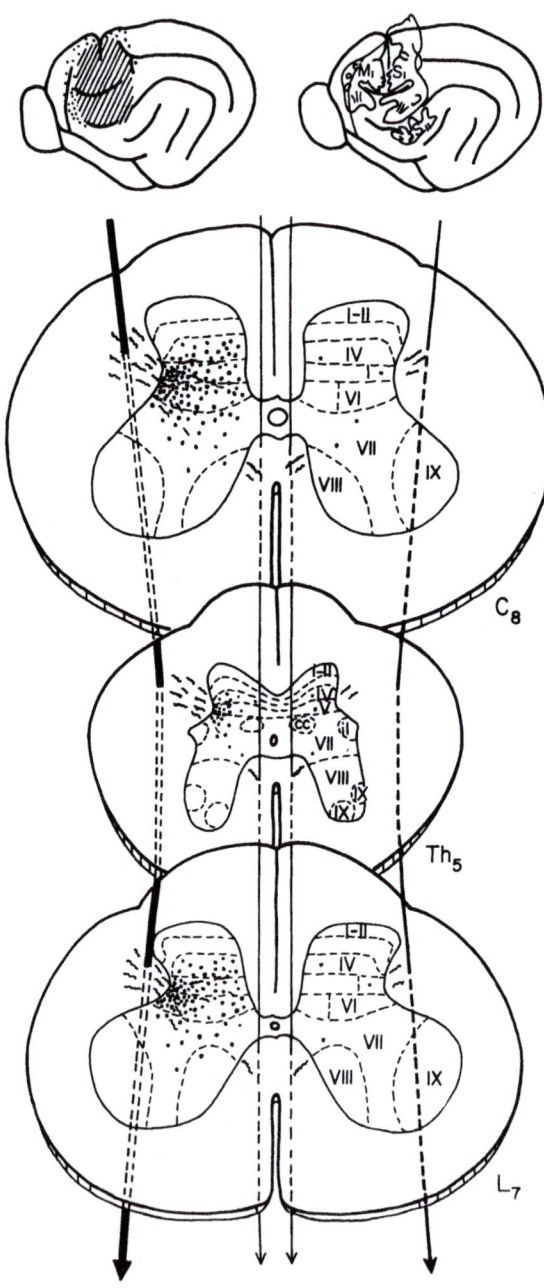

Fig. 1

some subcortical white matter and encroaches upon the head of the caudate nucleus. Degenerating corticospinal fibres descend bilaterally in the lateral and the ventral funiculi, even to the lowest lumbar segments. However, the uncrossed lateral and both ventral tracts are relatively small, especially in the lower cord segments, where the degenerating fibres can be identified only in longitudinal sections.

In the cervical and lumbar enlargements *the area of termination* is restricted to REXED's laminae IV, V, VI and the dorsal regions of lamina VII, except for an occasional degenerating fibre in lamina VIII. The fibres in the lateral corticospinal tract enter the spinal grey matter from its lateral aspect in laminae V—VI and spread out in a fan-shaped fashion. However, a dorsomedial and a ventromedial component can be distinguished. Fibres of the former are distributed largely to laminae IV—V, some to the dorsomedial parts of lamina VI. Many of these fibres acquire a longitudinal course within the grey matter. The fibres of the ventromedial component supply mainly lamina VI and the dorsal parts of lamina VII. The fibres of the ventral tract enter the grey matter from its medial aspect in lamina VIII, but on account of some intermingling with the laterally derived fibres it cannot be decided whether they have any preferential site of termination, although on the whole they appear to end largely ventromedially in lamina VII. No terminations are found in lamina IX harbouring the ventral motor horn cells.

In the thoracic cord the distribution of corticospinal fibres is essentially the same as in the enlargements, except for the differences due to the absence of lamina VI. No fibres have been found to end in the column of Clarke or in the intermediolateral cell column.

With regard to the *mode of termination* of corticospinal fibres, contacts by means of "terminal degeneration" have been found on the perikarya as well as on the dendrites of small as well as medium sized and large neurons within the laminae of termination (figs. 14—18).

The finding that in the cat no corticospinal fibres terminate directly on the ventral motor horn cells agrees with the results of silver impregnation studies of

Fig. 1. Diagrammatic representation of the distribution of degenerating coarser (wavy lines) and terminal fibres (dots) as seen in transverse sections from three representative levels of the spinal cord of a cat following a total lesion of the left primary sensorimotor cortex (above to the left). The extent of the lesion as seen on the surface of the hemisphere is hatched, the dots outline the area, the fibres from which have been interrupted. Above to the right, diagram of the cerebral cortex of the cat showing the cortical localization of the "motor" and "sensory" areas according to WOOLSEY (1958). The Roman numerals refer to the laminae of REXED (1952, 1954). *Abbreviations for all figures.* B.c.: brachium conjunctivum; Br.p.: brachium pontis; CC: column of Clarke; Ce: nucleus cervicalis centralis; C.g.m.: central grey matter; Co: nucleus commissuralis; C.p.: posterior commissure; C.r.: restiform body; C.s.: superior colliculus; C.t.: trapezoid body; D: descending (inferior) vestibular nucleus; F.l.m.: fasciculus longitudinalis medialis; F.r.: fasciculus retroflexus; il: intermediolateral cell column; im: intermediomedial nucleus; L: lateral vestibular nucleus of Deiters; L.m.: medial lemniscus; M: medial vestibular nucleus; MI: primary "motor" cortex; N.c.m.: nucleus of the mamillary body; N.c.p.: nucleus of the posterior commissure; N.c.t.: nucleus of trapezoid body; N.coll.inf.: nucleus of the inferior colliculus; N.cu.e.: external cuneate nucleus; N.D.: nucleus of Darkschewitsch; N.E.W.: nucleus of Edinger and Westphal; N.i.: nucleus interstitialis of Cajal; N.l.l.: nuclei of lateral lemniscus; N.m.X: dorsal motor (parasympatic) nucleus of vagus; N.mes.V: mesencephalic trigeminal nucleus; N.n.III, N.n.V, N.n.VI and N.n.VII: nuclei of III., V., VI. and VII. cranial nerves; N.III, N.V, N.VI and N. VII: III., V., VI. and VII. cranial nerves; N.pr.V.: principal sensory trigeminal nucleus; N.r.: nucleus ruber; N.r.l.: lateral reticular nucleus (nucleus of lateral funiculus); N.r.t.: nucleus reticularis tegmenti pontis; N.tr.s.: nucleus of solitary tract; N.tr.sp.V: nucleus of spinal trigeminal tract; Ol.i. and Ol.s.: inferior and superior olive; P: pontine nuclei; P.c.: cerebral peduncle; P.g.: periaqueductal grey matter; p.h.: nucleus praepositus hypoglossi; R.gc.: nucleus reticularis gigantocellularis; R.p.c.: nucleus reticularis pontis caudalis; R.p.o.: nucleus reticularis pontis oralis; S: superior vestibular nucleus; SI: primary "sensory" cortex; SII: secondary somatic area; S.n.: substantia nigra; tr.s.: solitary tract; tr.sp.V: spinal trigeminal tract; x: small-celled group x, lateral to the descending vestibular nucleus

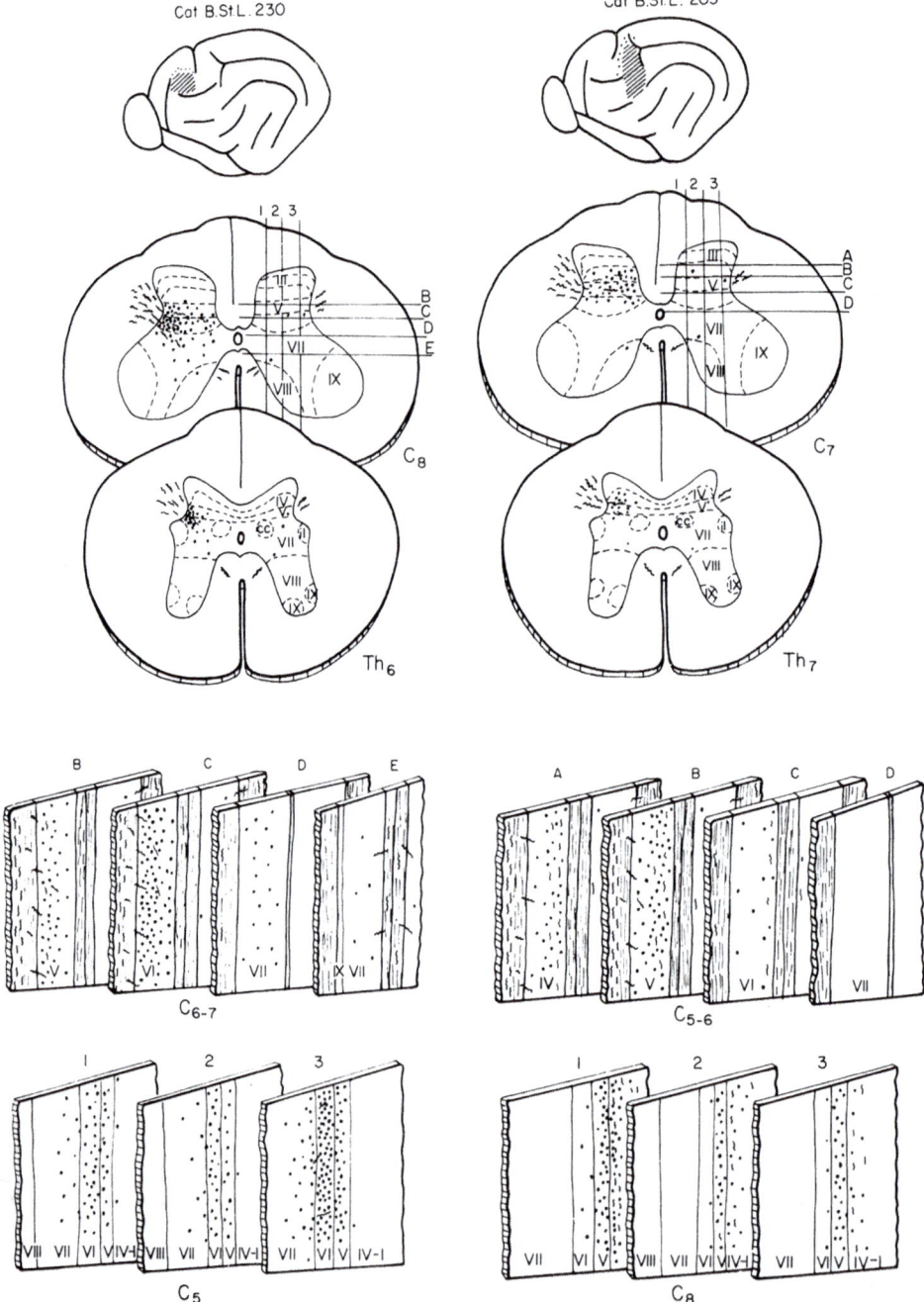

Fig. 2. Diagrammatic representation of the difference in distribution of degenerating terminal fibres (dots) within the cervical and thoracic grey matter in two cases with relatively isolated lesions of the forelimb "motor" and "sensory" cortices, respectively. Above, the extent of the lesions as seen on the surface of the hemisphere. From the thoracic cord only transverse sections are shown, while transverse as well as horizontal and sagittal sections are shown from the cervical enlargement. The positions of the horizontal and sagittal sections are indicated in the drawings of the transverse sections. The white matter is not shown in the sagittal sections. Symbols and abbreviations as in fig. 1

SZENTÁGOTHAI-SCHIMERT (1941) and CHAMBERS and LIU (1957) in the cat. How-ever, in the monkey KUYPERS (1960) and LIU and CHAMBERS (1964) report such direct connections.

Within the corticospinal fibre system there are interesting differences con-cerning the terminal regions in the spinal grey matter between cases with lesions of the "motor" or "sensory" parts of the primary sensorimotor area of the cerebral cortex. One pair of cases illustrating this is shown in fig. 2.

In cat B.St.L. 230 (killed after 7 days) the lesion is largely confined to the forelimb "motor" region. This case exhibits a small dorsomedial component of degenerating corticospinal fibres within the spinal grey matter. Its few fibres supply only lamina V laterally and the most ventral part of lamina IV. The ventromedial component is massively degenerated and distributes fibres to lamina VI, especially laterally, and the dorsal part of lamina VII.

Cat B.St.L. 203 (killed after 9 days) is an example of cases with lesions involv-ing mainly the primary "sensory" forelimb area. In this case the ventromedial component of corticospinal fibres is almost entirely absent. "Terminal" degenera-tion is present medially in laminae IV—V and in the dorsomedial parts of lamina VI, and appears chiefly to be derived from the dorsomedial component. These fibres appear to be thinner than those of the ventromedial component described in the case with lesion of the "motor" area. Corresponding distributions of degeneration are found in cases with lesions of the hindlimb regions of the "motor" and "sensory" areas, respectively.

These findings appear to agree with those made by KUYPERS (1960) in the macaque monkey. He reports corticospinal fibres coming from the precentral gyrus to terminate mainly in the "basal parts of the dorsal horn and the inter-mediate zone", while those from the postcentral gyrus terminate in the nucleus proprius of the dorsal horn, which according to REXED (1952) appears to corre-spond in part to lamina IV.

It should, however, be noted that in spite of the preferential distribution of fibres from the "motor" cortex to the ventrolateral regions, of those from the "sensory" cortex to the dorsomedial regions, there is no complete segregation. Thus both groups of corticospinal fibres appear to terminate in lamina V and the dorsomedial parts of lamina VI, but the "sensory" cortex seems to be almost alone in supplying lamina IV, while on the other hand, the "motor" cortex appears to be the only source of corticospinal fibres to lamina VII and the ventrolateral parts of lamina VI. Correspondingly most fibres in the ventromedial component of entering corticospinal fibres come from the "motor" cortex, while those from the "sensory" cortex appear to pass mainly in the dorsomedial component.

b) The rubrospinal fibre system

The red nucleus of mammals is generally considered to be composed of three types of nerve cells: large, medium sized and small ones. Traditionally the red nucleus has been divided into a caudal magnocellular part — consisting of large cells — and a rostral parvicellular part — composed only of small neurons. Recent studies of BRODAL and GOGSTAD (1954) and POMPEIANO and BRODAL (1957a), however, leave no doubt that in the cat even the most caudal part of the nucleus harbours some small nerve cells, and that the rostral "reticular" part, where the small neurons dominate, contains some large neurons. POMPEIANO and BRODAL (1957a) further demonstrated that the rubrospinal tract is somatotopically organized (fig. 3), and that neurons of all sizes from the whole red nucleus contribute fibres to the rubrospinal projection in the cat.

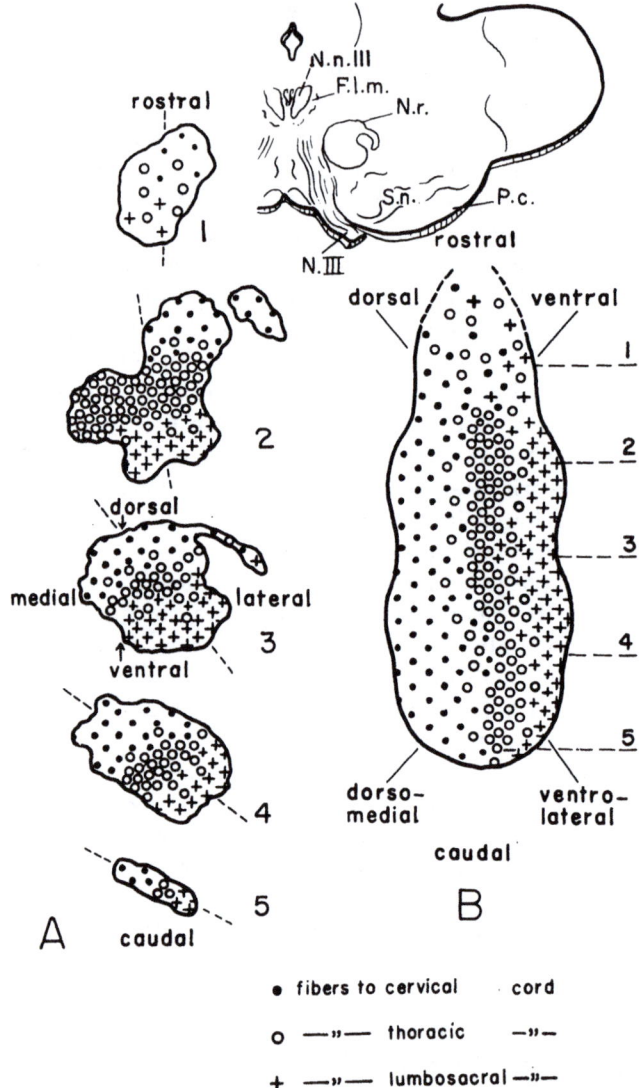

Fig. 3. Diagram showing the somatotopical pattern within the rubrospinal projection in the cat as demonstrated experimentally by POMPEIANO and BRODAL (1957a). To the left a series of transverse sections through the red nucleus, to the right a longitudinal reconstruction of the nucleus. Abbreviations as in fig. 1. From POMPIANO and BRODAL (1957a)

Fig. 4 shows the findings in a case from the study of NYBERG-HANSEN and BRODAL (1964). The lesion in cat B.St.L. 278 (killed after 9 days) is mainly restricted to the "forelimb" region of the red nucleus, according to the schema of POMPEIANO and BRODAL (1957a) presented in fig. 3. Degenerating rubrospinal fibres are seen to descend in the contralateral lateral funiculus even to the lowest segments of the lumbosacral enlargement. In the cervical cord the rubrospinal fibres are located in the ventral part of the dorsal half of the lateral funiculus, while during the descent in the thoracic cord they are gradually displaced in a dorsolateral direction and are finally located peripherally in the dorsolateral part

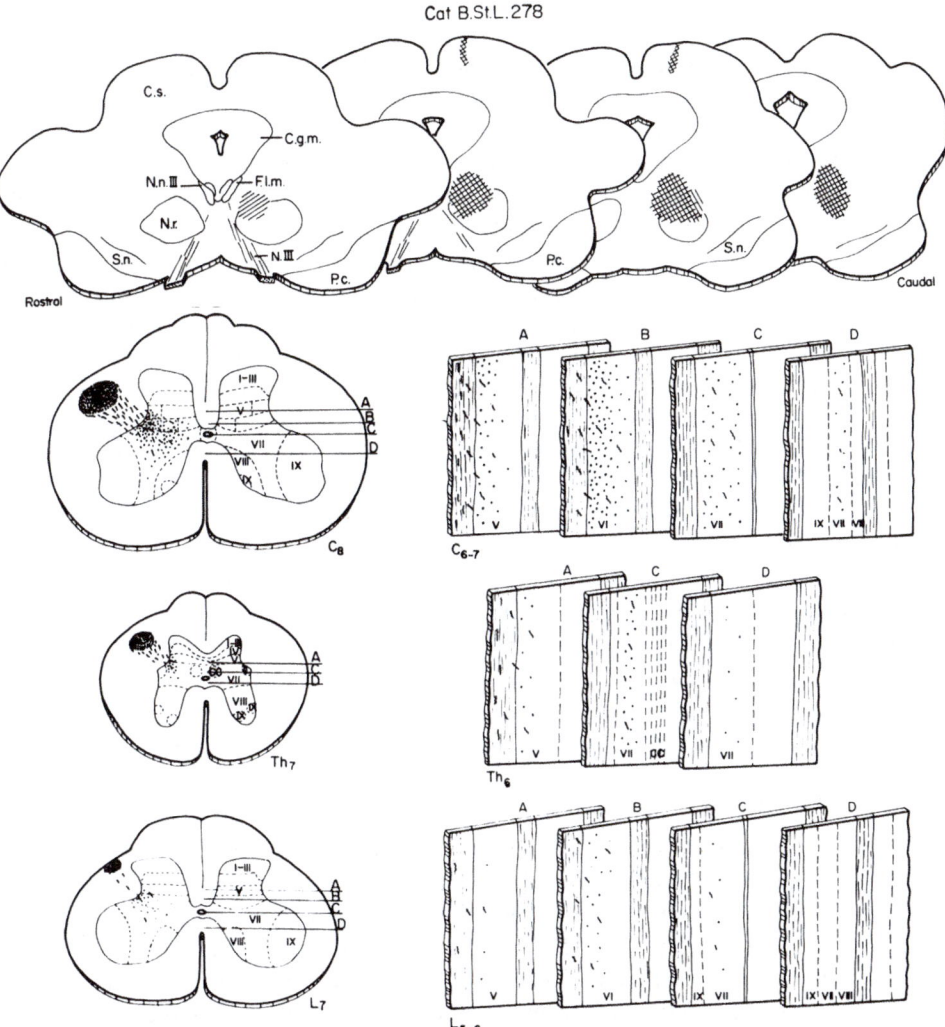

Fig. 4. Diagrammatic representation of the distribution of degeneration within the spinal cord as seen in transverse and horizontal sections from the cervical, thoracic and lumbar cord in a cat, having a lesion (hatching) mainly of the dorsomedial "forelimb" region of the red nucleus (above). Symbols and abbreviations as in fig. 1

of the lateral funiculus close to the most dorsal part of the dorsal horn in the lumbosacral enlargement. The observation that the rubrospinal tract in the cat descends even to sacral levels of the cord is in agreement with recent findings obtained by means of silver impregnation techniques (SZENTÁGOTHAI-SCHIMERT 1941; POMPEIANO and BRODAL 1957a; HINMAN and CARPENTER 1959; STAAL 1961) and the findings of POMPEIANO and BRODAL (1957a) using the modified

Gudden method (Brodal 1940). Kuypers et al. (1962) have made similar observations in the monkey.

The sites of entrance of degenerating rubrospinal fibres in the spinal grey matter correspond to the lateral parts of laminae V—VII, with a maximum ventrally in lamina VI (in the thoracic cord segments studied this lamina is absent). In cat B.St.L. 278 with the lesion located mainly in the dorsomedial "forelimb" region of the red nucleus, most of the degenerating fibres enter the spinal grey matter in the upper half of the cord. In cases in which the lesion is located in the ventrolateral "hindlimb" region of the red nucleus the bulk of fibres enters the grey matter in the lower half of the cord. These findings thus confirm the somatotopical pattern within the rubrospinal projection first demonstrated by Pompeiano and Brodal (1957a).

Having entered the spinal grey matter, the rubrospinal fibres radiate ventromedially in a fan-shaped fashion and spread to the lateral half of lamina V, the whole of lamina VI, although mainly laterally, and to lamina VII, especially laterally and centrally. No terminations are found on ventral motor horn cells in lamina IX. The latter observation confirms the findings of Rasdolsky (1923) using his "Licht-grün fuchsin" method in the dog and recent silver impregnation studies in the cat (Szentágothai-Schimert 1941; Staal 1961) and monkey (Kuypers et al. 1962).

In the thoracic cord the distribution of rubrospinal fibres is essentially the same as in the enlargements, except for the absence of lamina VI. No terminations are found on cells in the column of Clarke or the intermediolateral cell column. From a quantitative point of view, it appears that most rubrospinal fibres terminate in the enlargements, while the thoracic cord receives a smaller number.

Concerning the *mode of termination* of rubrospinal fibres, both Nauta and Glees sections demonstrate that the fibres end on the somata as well as along the dendrites of nerve cells of all sizes within the laminae of termination (figs. 19—22).

c) The lateral and medial vestibulospinal fibre systems

The projections from the vestibular nuclei to the spinal cord have been the subject of numerous studies. In spite of this considerable confusion exists in the literature, due to, among other things, inconsistencies in the nomenclature of the nuclear subgroups of the vestibular complex and the difficulties in producing lesions restricted to particular nuclei. In recent years various afferent and efferent connections of the vestibular nuclei have been studied in the Anatomical Institute in the University of Oslo (for reviews, see Brodal, Pompeiano and Walberg 1962; Brodal 1963a, 1964). The findings lend support to the correctness of the subdivision of the vestibular nuclei made on a cytoarchitectonic basis by Brodal and Pompeiano (1957).

In addition to the four classical nuclei, the superior, lateral, medial and descending (inferior), these authors identified several smaller cell groups. Only the lateral and medial vestibular nucleus concern us here, since, as will be shown below, only these two nuclei give origin to fibres descending to the spinal cord.

The *lateral vestibular nucleus* in addition to the giant cells of Deiters, contains a considerable number of medium sized and small neurons. The giant cells are generally more numerous and larger in the dorsocaudal than in the rostroventral part of the nucleus.

The *medial vestibular nucleus* is mostly composed of medium sized nerve cells, intermingled with some small neurons.

The fibre connections from the vestibular nuclei to the spinal cord are usually described as consisting of two separate fibre systems: the classical vestibulospinal tract and the fibres descending in the medial longitudinal fasciculus. For reasons to be given below, these two pathways will be called the lateral and medial vestibulospinal tract, respectively, in the present account.

The observation of POMPEIANO and BRODAL (1957 b) that the lateral vestibulospinal tract takes its origin only from the lateral vestibular nucleus has been confirmed by NYBERG-HANSEN and MASCITTI (1964). Lesions of the medial and descending nucleus do not result in degenerating fibres in the lateral vestibulospinal tract. There is no conclusive evidence that the superior nucleus projects to the spinal cord (POMPEIANO and BRODAL 1957 b).

Fig. 5 shows the findings in one of the cases (NYBERG-HANSEN and MASCITTI 1964) in which the lesion only involves the lateral nucleus (cat B.St.L. 307, killed after 8 days).

Degenerating lateral vestibulospinal fibres are seen in the ventral funiculus of the spinal cord on the homolateral side only. They descend the whole cord throughout. In the cervical cord they are situated peripherally in the ventrolateral funiculus and do not extend into the dorsal three-fourth of the ventral funiculus where the medial longitudinal fasciculus is located. During their descent in the thoracic cord, however, the fibres are gradually displaced in a dorsomedial direction, and in the lumbosacral enlargement they are found medially in the ventral funiculus along the anterior median fissure.

The degenerated lateral vestibulospinal fibres enter the spinal grey matter corresponding to the medial aspect of lamina VIII and radiate into other parts of the grey matter (figs. 23—24). There is pronounced "terminal" degeneration in the entire lamina VIII and in the neighbouring medial and central parts of lamina VII. In the thoracic cord segments where lamina VIII extends across the ventral horn, degenerating terminal fibres are not only found medially in lamina VIII, but in its lateral parts as well, although the density of degeneration is greatest medially.

At all levels of the cord entering degenerating lateral vestibulospinal fibres can be seen to by-pass the ventromedial group of motoneurons, and in the enlargements they have never been seen to establish synaptic contacts with these cells. In the thoracic cord, however, a few terminations are observed on the motoneurons in the ventromedial group in some sections. Although the area of termination reported by SCHIMERT (1938) appear to be in accord with the present findings, his statement of terminations almost exclusively on medial motoneurons is scarcely tenable, since the majority of the neurons in lamina VIII, by SCHIMERT interpreted as medial motoneurons, probably are propriospinal neurons (see REXED 1952; and NYBERG-HANSEN and MASCITTI 1964). In the present study, furthermore, terminations on the motoneurons in the larger lateral lamina IX have never been observed neither in the enlargements nor in the thoracic cord. This latter *observation confirms the findings of* RASDOLSKY (1923) *in the dog, and as far as*

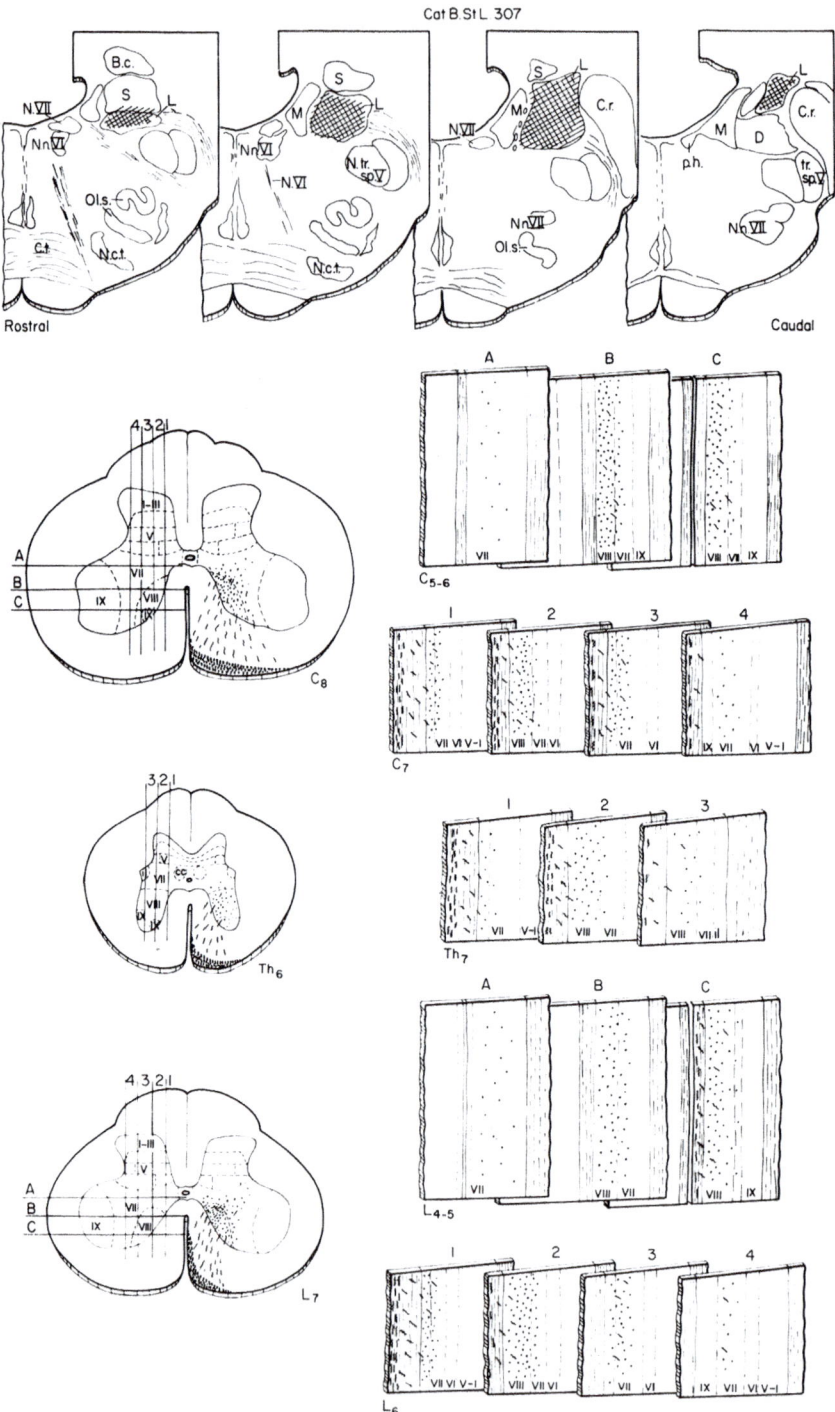

Fig. 5. Diagrammatic representation of the distribution of degenerating coarser (wavy lines) and terminal fibres (dots) within the spinal cord of a cat with a total lesion (hatchings) of the lateral vestibular nucleus of Deiters (above). From the thoracic cord only transverse and sagittal sections are shown, while horizontal sections as well are shown from the cervical and lumbar enlargements. Symbols and abbreviations as in fig. 1

can be judged from their papers KUYPERS et al. (1962) in the monkey and STAAL (1961) in the cat have made similar observations with silver techniques. Degenerating fibres have likewise never been found in the intermediolateral cell column or in the column of Clarke.

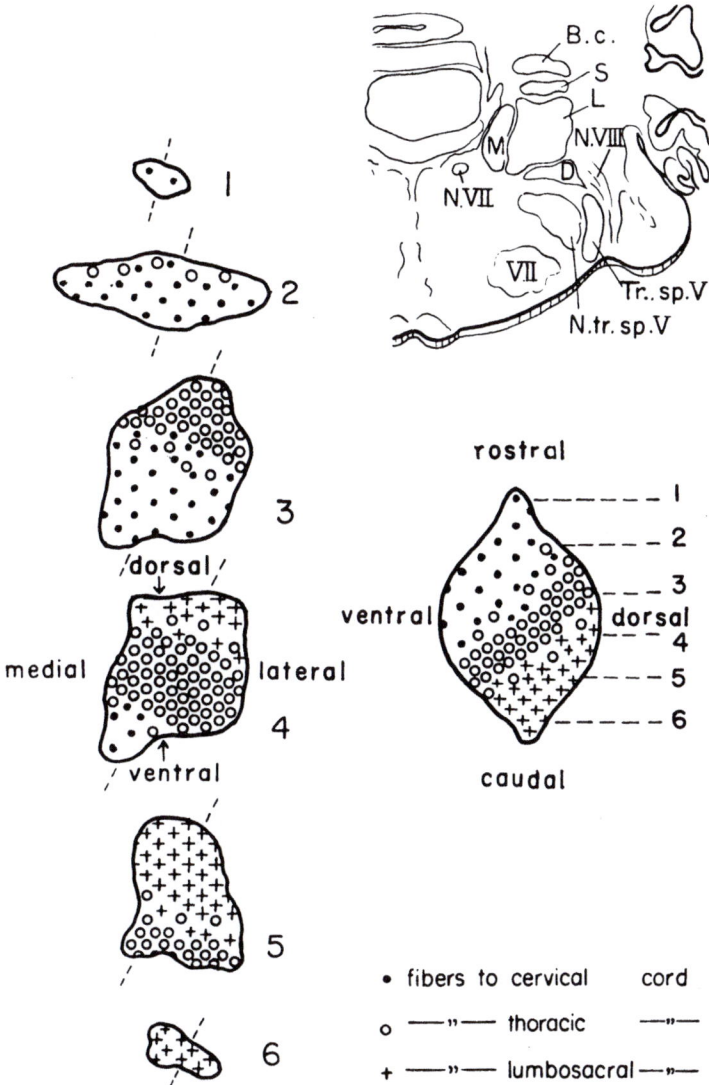

Fig. 6. Diagram showing the somatotopical pattern within the lateral vestibulospinal projection in the cat as demonstrated experimentally by POMPEIANO and BRODAL (1957b). To the left a series of transverse sections through the lateral vestibular nucleus, to the right a longitudinal reconstruction of the nucleus. Abbreviations as in fig. 1. From POMPEIANO and BRODAL (1957b)

From a quantitative point of view, the thoracic cord receives fewer fibres than the enlargements.

Observations made in another case confirm the somatotopic origin of the lateral vestibulospinal tract first demonstrated by POMPEIANO and BRODAL (1957b). Fig. 6 which diagramatically summarizes their findings, shows how the rostroventral part of the lateral nucleus projects to the cervical cord, while the fibres to the lumbosacral enlargement take origin from the dorsocaudal part. Fibres to the thoracic cord come from intermediate regions.

Concerning the *mode of termination,* the lateral vestibulospinal fibres terminate on nerve cells of all sizes within the laminae of termination. In the NAUTA sections it is seen that the majority of the fibres appear to establish axo-dendritic contacts

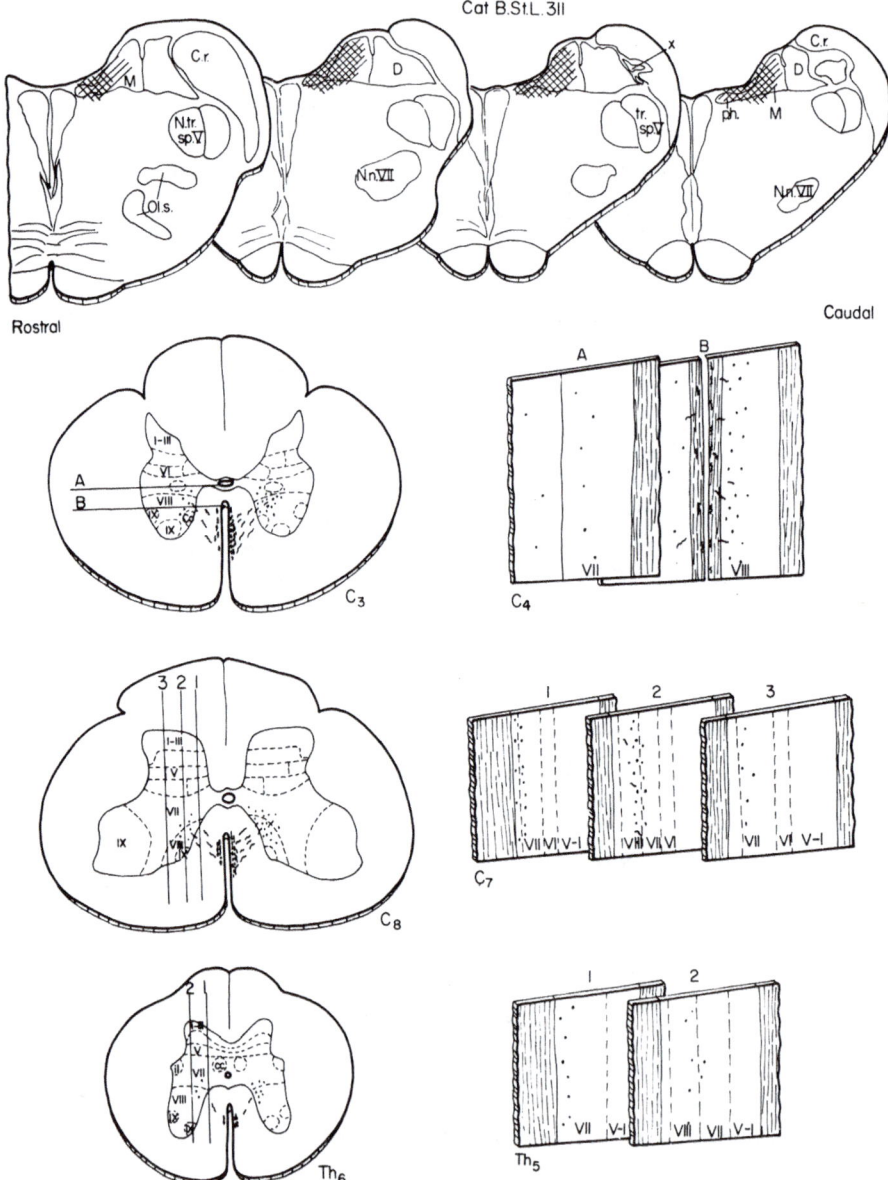

Fig. 7. Diagrammatic representation of the distribution of degenerating coarser (wavy lines) and terminal fibres (dots) within the spinal cord of a cat with a lesion (hatchings) involving the medial vestibular nucleus (above). From the thoracic cord only transverse and sagittal sections are shown, while horizontal sections as well are shown from the cervical cord. Symbols and abbreviations as in fig. 1

with the proximal parts of the cell processes, while only a small number terminate on the somata. The GLEES sections in general confirm these findings, although the latter method reveals more axo-somatic contacts than does the NAUTA method (figs. 25—28).

The *medial vestibulospinal tract* (NYBERG-HANSEN 1964a) is shown in fig. 7, which demonstrates the observations in cat B.St.L. 311 (killed after 7 days) in which the lesion is located to the medial vestibular nucleus. Degenerating vestibulospinal fibres are found in the dorsal three-fourths of the ventral funiculus of the spinal cord within the area along the anterior median fissure generally alotted to the medial longitudinal fasciculus. On account of their origin in the medial nucleus and their medial course, it is proposed that these fibres should be called *medial vestibulospinal fibres*, while those from the lateral vestibular nucleus coursing more laterally outside the medial longitudinal fasciculus should be called lateral vestibulospinal fibres[1].

The medial vestibulospinal fibres can be traced bilaterally downwards to the midthoracic cord segments. The fibres on the homolateral side outnumber those on the contralateral. These findings seem to be in accordance with the older observations of VAN DER SCHUEREN (1912), BUCHANAN (1937) and FERRARO, PACELLA and BARRERA (1940). In several cases with lesions in the superior, lateral and descending vestibular nucleus, no degeneration was observed in the medial vestibulospinal tract (NYBERG-HANSEN 1964a).

Within the spinal grey matter the medial vestibulospinal fibres terminate bilaterally in the dorsal half of lamina VIII and the neighbouring medial parts of lamina VII in the upper half of the cord (fig. 29). Their number is, however, much smaller than the lateral vestibulospinal fibres, which are to be regarded as making up the major pathway from the vestibular nuclear complex to the spinal cord.

d) The reticulospinal fibre systems

According to BRODAL (1957) the feline reticular formation (in the following abbreviated to RF) may be divided into two major parts: the medial two-thirds approximately are characterized by neurons of all sizes, small, medium sized and large ones, some even of giant size; the lateral third is made up of small nerve cells intermingled with some medium sized ones. The existence of fibres taking origin in various areas of the RF and descending to the spinal cord has been known since the paper of TSCHERMAK (1898). Since then several authors have dealt with the origin, course and even termination of reticulospinal fibres. The data obtained, however, are incomplete and conflicting on many points. The most extensive and complete study has been published by TORVIK and BRODAL (1957) who, using the modified Gudden method (BRODAL 1940) of retrograde degeneration, defined the sites of origin of reticulospinal fibres from the pons and from the medulla. The

---

[1] The term descending medial longitudinal fasciculus should not be used in connection with spinal projections from the vestibular nuclei since in the cord it only denotes the dorsal three-fourth of the ventral funiculus along the anterior median fissure. As will be shown in the following sections, interstitiospinal and pontine reticulospinal fibres as well are situated in this region in the spinal cord. By calling the spinal projection from the medial vestibular nucleus coursing within the descending medial longitudinal fasciculus medial vestibulospinal fibres confusion on this point can be avoided.

former were found to descend in the ventral funiculus of the cord, while those from the medulla oblongata descend in the ventral half of the lateral funiculus on both sides.

Taking advantage of these findings NYBERG-HANSEN (1965a) made discrete lesions in different areas of the brain stem RF. Following lesions of the mesencephalic RF, it was not possible to observe degenerating fibres in the spinal cord, and the conclusion was reached that this part of the RF does not send fibres directly to the spinal cord in the cat.

Fig. 8a shows the findings in cat B.St.L. 299 (killed after 8 days) in which the lesion involves large parts of the *nucleus reticularis pontis caudalis*, where TORVIK and BRODAL (1957) found the majority of pontine reticulospinal fibres to originate, with concomitant destruction of the nucleus reticularis tegmenti pontis (which does not project to the cord) and the medial part of the medial lemniscus. Other fibre systems coursing through the pontine RF are not touched upon.

Within the spinal white matter degenerating pontine reticulospinal fibres are almost exclusively homolateral and descend in the entire ventral funiculus. The most dorsomedial fibres are located within the area generally allotted the descending medial longitudinal fasciculus in the cord (see footnote [1]), while the most lateral fibres reach the ventral root fibres emerging most laterally. The pontine reticulospinal fibres descend even to the lowest segments of the lumbosacral enlargement, and the majority of them appear to terminate in the enlargements. Some of the most dorsomedially located fibres in the homolateral ventral funiculus cross in the anterior commissure at all spinal levels and terminate in the grey matter on the contralateral side together with the few pontine reticulospinal fibres which descend in the contralateral ventral funiculus. The observation that pontine reticulospinal fibres descend even to lumbosacral levels is in agreement with the old findings of TSCHERMAK (1898) and the recent studies with silver techniques of STAAL (1961) in the cat and KUYPERS et al. (1962) in the monkey.

Degenerating pontine reticulospinal fibres enter the spinal grey matter corresponding to lamina VIII. A few fibres enter lamina VII and the ventral parts of lamina IX. Having entered, the degenerating fibres radiate further into the grey matter and terminate in the entire lamina VIII and the medial neighbouring parts of lamina VII. In the thoracic cord where lamina VIII extends across the ventral horn, degenerating pontine reticulospinal fibres are found not only medially in lamina VIII, but in its lateral regions as well. Nearly all of the fibres which enter lamina IX appear to radiate through this lamina and to terminate in lamina VII. In some sections a few fibres can, however, be seen to terminate within the ventromedial group of motoneurons. Neither STAAL (1961) nor KUYPERS et al. (1962) found evidence for direct terminations of pontine reticulospinal fibres on ventral motor horn cells.

Fig. 8b shows the finding in cat B.St.L. 318 (killed after 7 days) with a lesion in the *nucleus reticularis gigantocellularis* with some destruction of the dorsomedial part of the inferior olive. The lesion thus corresponds to the areas of the RF which according to TORVIK and BRODAL (1957) give origin to the medullary reticulospinal fibres.

Within the spinal white matter degenerating descending fibres can be seen in both ventral funiculi, chiefly homolaterally, as described for cat B.St.L. 299

Cat B.St.L. 299

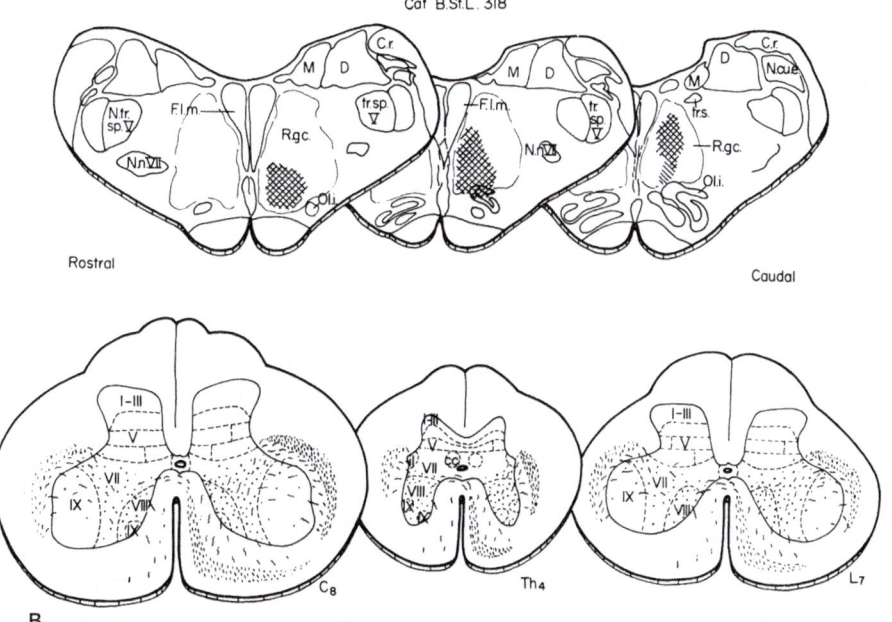

Fig. 8. Diagrammatic representation of the distribution of degenerating coarser (wavy lines) and terminal fibres (dots) as seen in transverse sections from three representative levels of the spinal cord in the cat following lesions of the nucleus reticularis pontis caudalis (A) and the nucleus reticularis gigantocellularis (B), respectively. Symbols and abbreviations as in fig. 1

(fig. 8a), and in the ventral parts of the lateral funiculi on both sides. It appears from a study of several cases (see NYBERG-HANSEN 1965a) that the degenerating fibres in the ventral funiculus are presumably derived from the pontine RF, because reticulospinal fibres from the pons pass through the medial longitudinal fasciculus and the RF lateral to this pathway in the lower brain stem, and are thus easily interrupted by lesions in the RF of the medulla.

On the basis of the findings of TORVIK and BRODAL (1957) the degenerating fibres in the lateral funiculus following lesions of the medullary RF, are considered

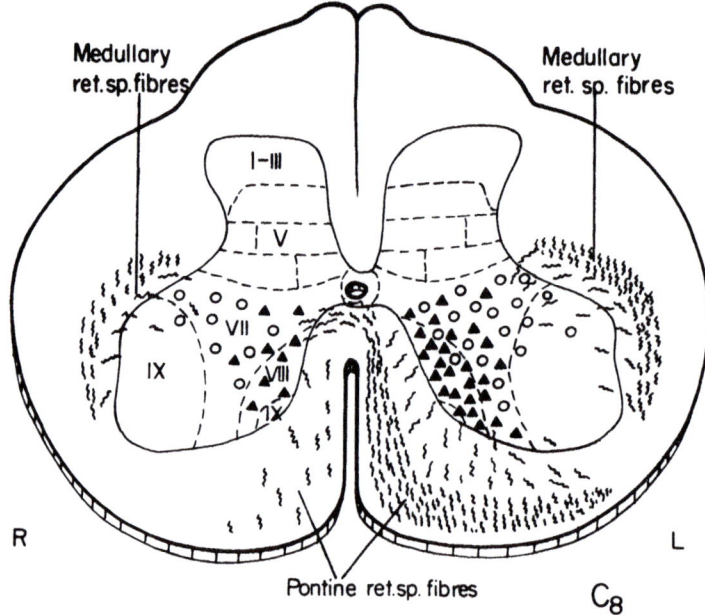

▲ Sites of termination of pontine ret. sp. fibres
○ Sites of termination of medullary ret. sp. fibres

Fig. 9. Diagram of a transverse section of the spinal cord at C₈, showing the location in the ventral and lateral funiculi, and the sites of termination within the spinal grey matter of the pontine (▲) and medullary (○) reticulospinal fibres arising in the nucleus reticularis pontis caudalis and nucleus reticularis gigantocellularis, respectively. Note the different location in the white matter, and the partially dissimilar areas of termination within the grey matter of the two contingents of reticulospinal fibres

as medullary reticulospinal fibres. Their number descreases as more caudal levels are reached, but the principal pattern outlined above is essentially similar at all spinal levels, even in the sacral cord. The fibres descending in the homolateral lateral funiculus outnumber those in the contralateral.

The medullary reticulospinal fibres enter the spinal grey matter corresponding to the ventrolateral part of lamina VII, while some enter the neighbouring parts of lamina IX and the ventral part of lamina VI. The entering fibres radiate ventrally in a fan-shaped fashion and terminate in the entire lamina VII, mainly in its central parts. The fibres entering lamina IX mostly appear to radiate through this and to terminate in lamina VII, while some terminate in lamina IX. Because of their small number it is difficult to obtain exact information as to the kind of neurons with which they establish synaptic contacts. A few are, however, seen in close relationship to some of the largest and some of the smallest nerve cells in lamina IX. No data are found in the literature giving specific information of the sites of termination of medullary reticulospinal fibres within the spinal grey matter. Fig. 9 summarizes the location in the white matter and the sites of termination in the grey matter of the pontine and medullary reticulospinal fibres.

As concerns the *mode of termination*, the relationships are essentially similar for pontine and medullary reticulospinal fibres. Thus, axo-somatic and axo-dendritic contacts are observed on large as well as on medium sized and small neurons within the laminae of termination (figs. 30—35). Furthermore, in the thoracic cord, neither the pontine nor the medullary reticulospinal fibres terminate in the column of Clarke or in the intermediolateral cell column.

e) The tectospinal fibre system

The feline superior colliculus exhibits a structural pattern of six principal layers which is found in most vertebrates. Most tectospinal fibres are usually reported to arise from the fifth layer, stratum griseum profundum, which is composed of medium sized triangular to pyramidal shaped neurons intermingled with some larger multipolar nerve cells.

Fig. 10 shows one of the cases from a study on the tectospinal projection in the cat (NYBERG-HANSEN 1965 b). The lesion mainly involves the superior colliculus. In the cord the degenerating tectospinal fibres are confined to the contralateral side and descend in the most ventral part of the ventral funiculus along the anterior median fissure as far down as the lowest segments of the cervical enlargement. The majority of the fibres terminate in the four upper cervical segments. This agrees with the findings of ALTMAN and CARPENTER (1961) and RASMUSSEN (1936).

The sites of entrance of the degenerating tectospinal fibres in the spinal grey matter of the ventral horn correspond to the ventromedial parts of lamina VIII, from which the fibres radiate and terminate in central and lateral parts of lamina VII, while some end in lamina VIII itself, and others proceed into the ventrolateral part of lamina VI. Although fibres can be seen to terminate in the neighbourhood of cells of the nucleus spinalis of the accessory nerve and the ventromedial group of motoneurons, no "terminal" degeneration is found on the motoneurons themselves, in contrast to what is mentioned in a brief abstract by PEARCE and GLEES (1956) and the recent findings of STAAL (1961). "Terminal" degenerating fibres were found neither in the nucleus commissuralis nor in the nucleus cervicalis centralis. The lack of direct terminations on the motoneurons agrees with the findings of RASDOLSKY (1923) and SZENTÁGOTHAI-SCHIMERT (1941), who report the tectospinal fibres to terminate in the so-called "intermediate zone".

With regard to the *mode of termination* of the degenerating tectospinal fibres, "preterminal" and "terminal" degenerating fibres are found on neurons of all sizes, large as well as medium sized and small ones within the laminae of termination (figs. 36—38).

Similar findings were obtained in two other animals with lesions in the superior colliculus, while no degenerating fibres at all could be found in the cord in one case with a lesion of the nucleus of the inferior colliculus.

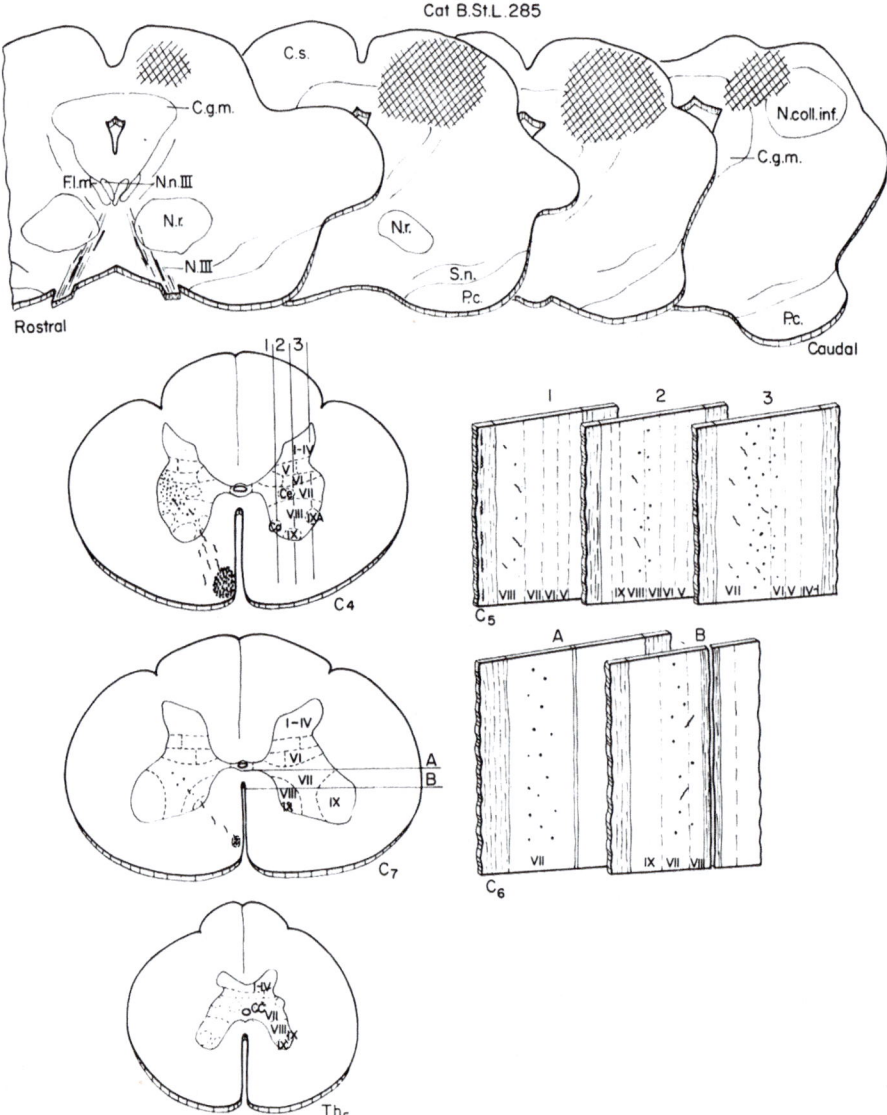

Fig. 10. Diagrammatic representation of the distribution of degenerating coarser (wavy lines) and terminal fibres (dots) within the spinal cord of a cat with an almost total lesion (hatchings) of the superior colliculus (above). From the thoracic cord only transverse sections are shown, while horizontal and sagittal sections as well are shown from the cervical cord. Symbols and abbreviations as in fig. 1

f) The interstitiospinal fibre system

The interstitial nucleus of Cajal, also called the nucleus of the medial longitudinal fasciculus, is situated close to the central grey matter in the rostral part of the mesencephalon. It has a reticular appearance and is composed of small and medium sized nerve cells, triangular to multipolar, intermingled with some larger multipolar neurons. Fibres coursing from the region of the interstitial nucleus to the spinal cord have been known since the publications of KOHNSTAMM (1900), PROBST (1900) and LEWANDOWSKY (1904), but since then little attention has been

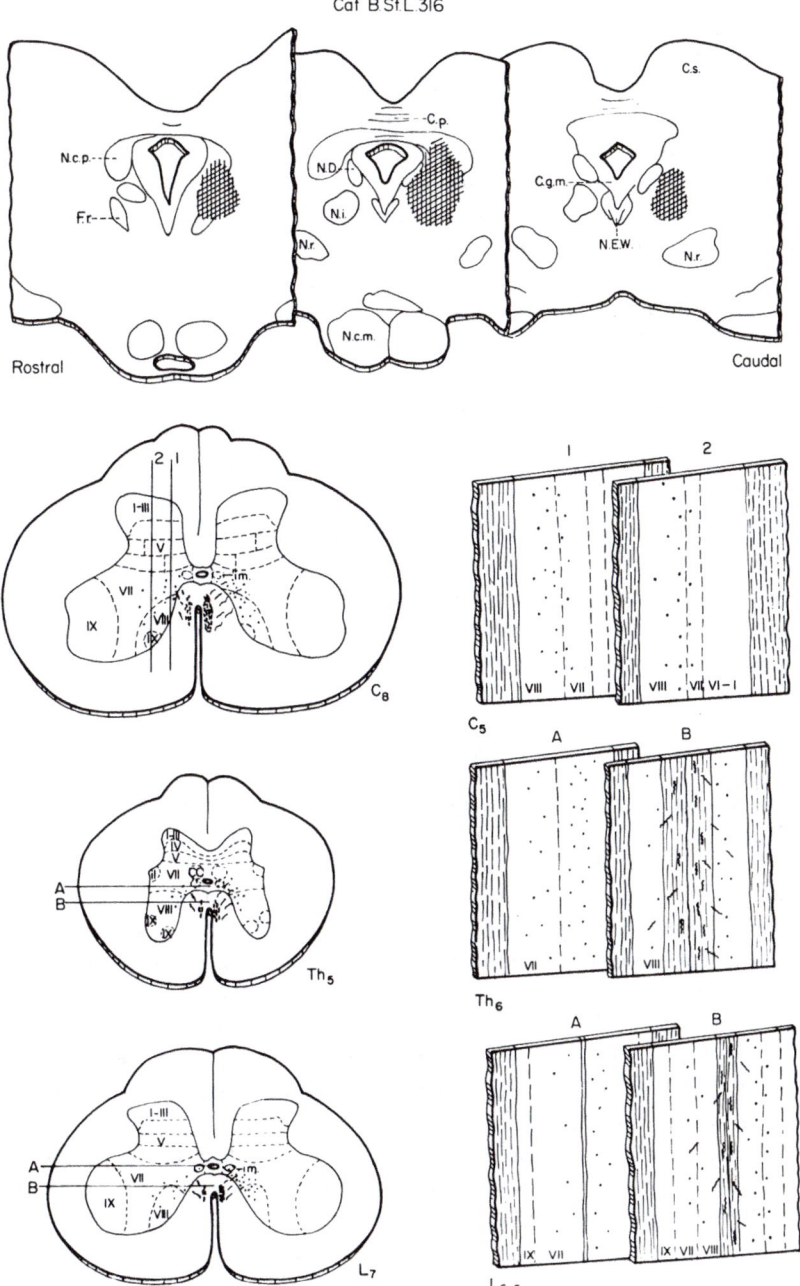

Fig. 11. Diagrammatic representation of the distribution of degenerating coarser (wavy lines) and terminal fibres (dots) within the spinal cord of a cat with a lesion (hatchings) involving the interstitial nucleus of CAJAL and parts of the nucleus of Darkschewitsch and the nucleus of the posterior commissure. From the thoracic and lumbar cord transverse and horizontal sections are shown, while transverse and sagittal sections are shown from the cervical cord. Symbols and abbreviations as in fig. 1

paid to these fibres. In connection with the study of the other descending fibre systems to the cord described above, it was deemed of interest to attempt a corresponding analysis of the interstitiospinal fibres. The main findings (NYBERG-HANSEN 1966, in press) will be briefly presented.

Fig. 11 shows the findings in cat B.St.L. 316 (killed after 5 days) in which the lesion involves mainly the interstitial nucleus, which is completely destroyed. Within the spinal cord descending degenerating fibres are observed in the dorsal part of the ventral funiculus on both sides, although the majority are found homolaterally. The intensity of degeneration decreases as more caudal levels are reached, but there are still some fibres left as far down as sacral segments. Recently STAAL (1961) has made similar findings.

In other cases where the lesions involve the entire nucleus of Darkschewitsch and the greater part of the nucleus of the posterior commissure, no degenerating fibres are found in the spinal cord. It may, therefore, be concluded that the degenerating fibres observed in the case shown in fig. 11 take their origin from the interstitial nucleus of Cajal, and thus are interstitiospinal fibres.

The degenerating interstitiospinal fibres enter the spinal grey matter dorsally in lamina VIII where some terminate, while others end in the neighbouring parts of lamina VII. No fibres terminate on ventral motor horn cells. Neither in the column of Clarke nor in the intermediolateral cell column are there degenerating fibres. A few appear, however, to terminate in relationship to neurons in the intermediomedial nucleus, especially in the thoracic and sacral cord (fig. 39). The sites of termination described above seem to concord with those reported by STAAL (1961) who, however, found degeneration in the homolateral spinal grey matter only.

IV. Discussion

As reviewed above, the descending fibre systems to the cord are not identical with regard to their sites and mode of ending in the spinal grey matter. Yet, some of them reveal striking mutual resemblances, while they differ from others. Morphological differences and similarities like these presumably reflect functional features. It seems a likely assumption, for example, that fibre systems ending in the same regions and in the same manner are functionally closely related to each other, while they presumably differ functionally from systems having different sites of termination. There are, in fact, a number of physiological observations from recent years which are in complete agreement with this assumption, as will be discussed below. These data achieve a broader interest when considered in relation to some other recent anatomical and physiological observations on those fibre systems which act on the nuclei of origin of the descending pathways to the cord. Finally, certain comparative anatomical features deserve mention. These will briefly be considered first.

a) Phylogenetical consideration

In cyclostomes, in which the first signs of cephalization within the central nervous system become apparent in phylogenesis (KAPPERS, HUBER and CROSBY 1936) neurons of the brain stem reticular formation provide the only pathway by which impulses from higher levels of the central nervous system reach the spinal cord. In these animals reticulospinal fibres thus furnish the final common path for descending supraspinal impulses to the cord. In plagiostomes in addition vestibulo-spinal fibres are developed. In reptiles tectospinal fibres begin to appear, and in birds they are well developed. Birds also exhibit a small rubrospinal tract. The latter, however, is much larger in mammals in which in addition a direct projection from the cerebral cortex to the spinal cord is established. This corticospinal fibre system undergoes a progressive enlargement as one passes from lower to higher mammals. Furthermore, the other descending fibre systems increase markedly in mammals as compared with lower forms. Consequently, in lower animals the ventral funiculus of the spinal cord where all the phylogenetically old supraspinal fibre systems descend is relatively much larger than the lateral funiculus containing the phylogenetically younger rubrospinal and corticospinal fibres.

The sequence of myelination in man of the fiber systems under consideration in general appears to parallel the order in which the fibers are differentiated phylogenetically. Thus the vestibulospinal fibres are among the first to acquire their myelin sheaths, while the corticospinal fibres are among the last.

b) Sites of termination of supraspinal fibres

In the light of the phylogenetical pattern outlined above the sites of termina-tion of descending supraspinal fibre systems as determined in the cat reveal some interesting features. The terminal regions of the pontine reticulospinal and lateral vestibulospinal fibres are very much the same (fig. 12 A). On the other hand, the sites of termination of the medullary reticulospinal fibres roughly coincide with the ventral terminal area of the rubrospinal and the extreme ventral terminal region of the corticospinal fibres (fig. 12 B).

It is interesting to notice that the three fibre systems descending in the lateral funiculus are spatially closely related to the laterally and dorsally localized spinal motoneurons which innervate the distal extremity musculature, while the fibre systems in the ventral funiculus, on the other hand, are spatially related to more ventromedially situated motoneurons innervating proximal muscles of the extremity. The latter muscles are mainly concerned with tonic extension move-ments, while the former to a large extent are engaged in more phasic flexion move-ments. On a pure morphological basis the suggestion may be ventured that the descending fibres in the lateral funiculus chiefly act on flexor motoneurons (via spinal interneurons), while those in the ventral funiculus mainly influence extensor motoneurons. These suggestions are indeed supported by recent physiological observations. Thus, the lateral vestibulospinal and the pontine reticulospinal fibres are known to convey impulses mainly facilitatory to extensor and inhibitory to flexor motoneurons (for references, see BRODAL, POMPEIANO and WALBERG 1962; and ROSSI and ZANCHETTI 1957, respectively). The corticospinal, rubro-spinal and medullary reticulospinal fibres chiefly exert just the opposite effects on the spinal mechanisms (see LUNDBERG and VOORHOEVE 1962; CORAZZA et al. 1963; PRESTON and WHITLOCK 1963; POMPEIANO 1957; SASAKI, NAMIKAWA and HASHIRAMOTO 1960; and ROSSI and ZANCHETTI 1957, respectively). It is indeed thoughtcompelling to notice that descending supraspinal fibre systems which exert similar influences on the spinal mechanisms have partly the same areas of

A

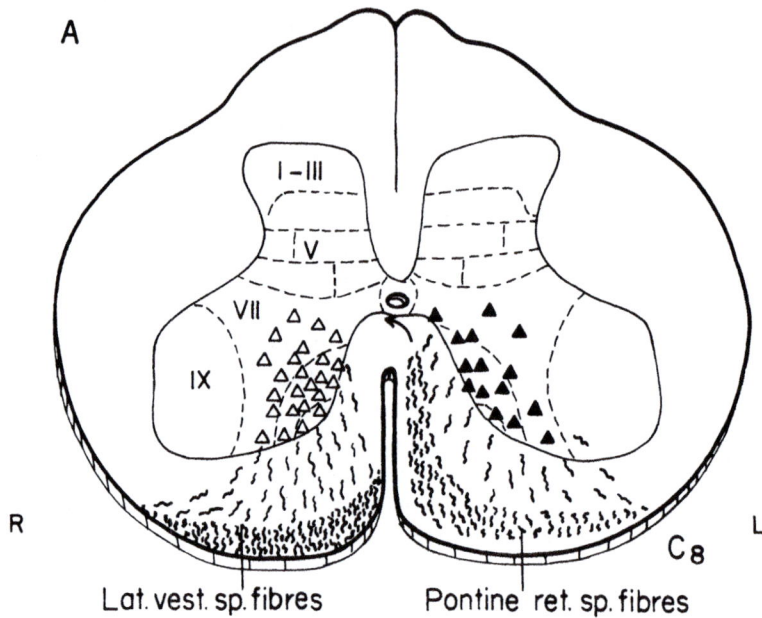

Lat. vest. sp. fibres   Pontine ret. sp. fibres

△ Sites of termination of lat. vest. sp. fibres
▲ Sites of termination of pontine ret. sp. fibres

B

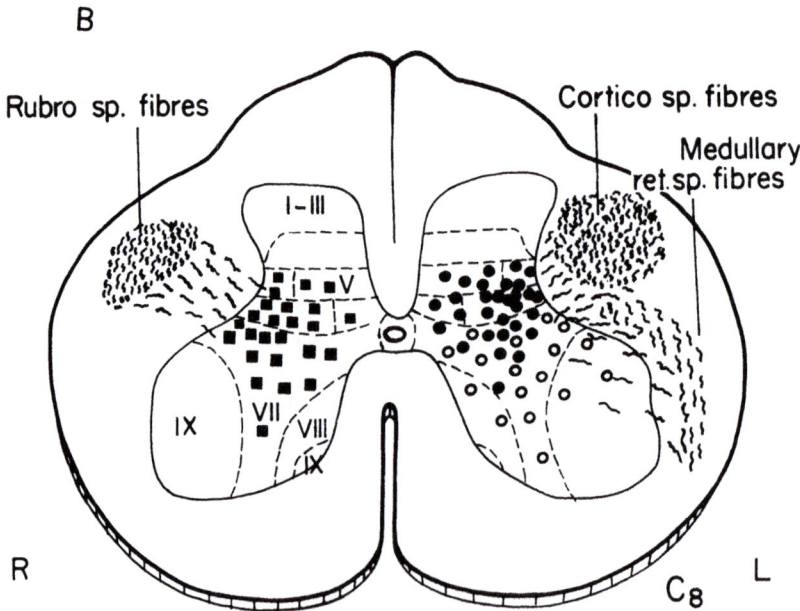

■ Sites of termination of rubrosp. fibres
● Sites of termination of corticosp. fibres from "motor" cortex
○ Sites of termination of medullary ret. sp. fibres

Fig. 12 A and B

termination within the spinal grey matter. Furthermore, the phylogenetically old supraspinal fibres, as pontine reticulospinal and vestibulospinal fibres, mainly descending in the ventral funiculus, exert excitatory influences on the spinal *extensor* mechanisms, and thus contribute to the maintenance of the postural tone. On the other hand, the phylogenetically younger rubrospinal and cortico-spinal fibre systems, descending in the lateral funiculus, chiefly exert facilitatory influences on the cord's *flexor* mechanisms, which from a physiological point of

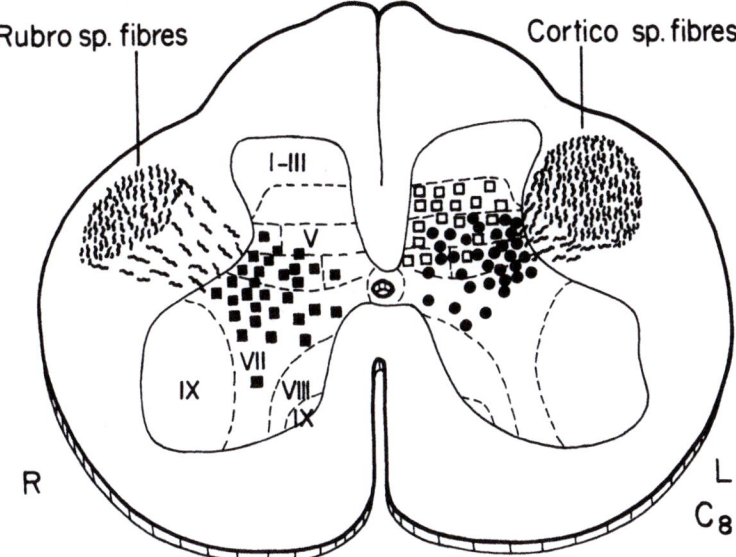

- ■ Sites of termination of rubro sp. fibres
- • Sites of termination of corticosp. fibres from "motor" cortex
- □ Sites of termination of corticosp. fibres from "sensory" cortex

Fig. 13. Diagram of a transverse section of the spinal cord at C_8, showing the location in the lateral funiculus and the sites of termination within the spinal grey matter of rubrospinal (■) and corticospinal fibres from "motor" (•) and "sensory" cortices (□), respectively. Note the similarities concerning the location in the white matter and the areas of termination within the grey matter of rubrospinal and corticospinal fibres from "motor" cortex

view develop more in phylogenesis than the extensor mechanisms and, further-more, provide the functional basis for the performance of so-called "skilled voluntary movements".

The clinical picture of spasticity has some bearing on this point. Spasticity is usually spoken of as a condition with increased muscle tone and exaggerated myotatic (deep or tendon) reflexes affecting the extensor (antigravity) muscles. It is, however, often forgotten that the opposite phenomena simultaneously occur in the flexor muscles. Spasticity thus should be considered as a condition with

Fig. 12 A and B. Diagrams of transverse sections of the spinal cord at C_8, showing the location in the white matter and the sites of termination within the spinal grey matter of lateral vestibulospinal (△) and pontine reticulospinal fibres (▲) in A, and of corticospinal from "motor" cortex (•), rubrospinal (■) and medullary reticulospinal fibres (○) in B. Note the similarities with regard to the location in the white matter, and the sites of termination within the grey matter of the two fibre systems in A on one hand, and of the three fibre systems in B on the other. Note, furthermore, the partial differences between the systems represented in each of the two diagrams

increased activity in extensor and decreased activity in the spinal flexor mechanisms. In accordance with the above account on the function of the various descending supraspinal fibre systems, spasticity is likely to occur when the activity in the systems in the ventral funiculus (lateral vestibulospinal and pontine reticulospinal fibres) is increased relatively, at the expense of those in the lateral funiculus (corticospinal, rubrospinal and medullary reticulospinal fibres). This may happen when the descending systems in the lateral funiculus themselves are affected in diseases of the spinal cord, or when afferent fibre connections to their structures of origin (for example fibres from the cerebral cortex and the cerebellum to the reticular formation and the red nucleus) are diseased or interrupted in pathological conditions at a supraspinal level.

c) Intrinsic organization of spinal mechanisms

The anatomical and physiological similarities and differences between the sites of termination of the various descending supraspinal fibre systems invite the suggestion that the reciprocal organization of motor performances may to a large extent depend on the intrinsic organization of the spinal cord itself. There is indeed physiological evidence that spinal interneurons of different functional types are intercalated 1: in a pathway excitatory to extensor and inhibitory to flexor motoneurons and 2: in a pathway mediating the reverse effect (see Eccles 1957; 1964, for recent reviews). The effect obtained on stimulation of a particular supraspinal structure or its descending fibres may then bear some relation to the kind of interneurons upon which the descending impulses converge. From a correlation between neurophysiological observations and anatomical data, it may be postulated that interneurons in pathways which are excitatory to flexor and inhibitory to extensor motoneurons are located within laminae V—VII, chiefly laterally. Interneurons of the flexion reflex pathway appear to have a similar location. On the other hand, interneurons mediating influences which are inhibitory to flexor and excitatory to extensor motoneurons of the spinal cord are chiefly located in lamina VIII and the adjacent medial parts of lamina VII. While the latter are found ventromedially, the former interneurons are found more dorsolaterally in the spinal grey matter (cp. fig. 12A and B). There is indeed some physiological evidence in favour of these assumptions. Thus, Bernhard and Rexed (1945) found interneurons mediating polysynaptic impulses to spinal flexor motoneurons to be localized chiefly in the lateral part of lamina VII, while those relaying activity to extensor motoneurons were inferred to be situated more ventromedially, corresponding to medial parts of lamina VII and the adjacent part of lamina VIII.

This line of reasoning concerning the intrinsic functional organization of the spinal cord is related to Bates' (1957) views, based among other things on the similarity between the movements which can be elicited on cortical stimulation in man and those which occur spontaneously in infants. The latter movements are, too a large extent believed to be organized at the spinal level. Sherrington (1906) stressed the resemblances between movements evoked by stimulation of the cerebral cortex in intact animals and those produced by sensory stimulation in spinal animals. The spinal mechanisms may thus be considered as a sort of claviature played upon by afferent impulses from many sources, among them supraspinal structures, which by way of descending fibre systems may act on and modulate more basic patterns of movements organized at the spinal level.

d) Descending supraspinal impulses

The extensive investigations of LUNDBERG's group (see LUNDBERG 1964b) of the mechanisms by which corticospinal impulses act at spinal levels are of relevance to the aspects just considered. They first demonstrated that the effects of corticospinal tract stimulation on the spinal motoneurons (mainly excitatory on flexor and inhibitory on extensor motoneurons), are secondary to the activation of interneurons of segmental reflex arcs (LUNDBERG and VOORHOEVE 1962). Furthermore, by intracellular recording from spinal interneurons, it was shown that the corticospinal fibres convey impulses which facilitate the inhibitory interneurons (AI) in the Ia inhibitory pathway, the excitatory (BE) and inhibitory (BI) interneurons in the Ib reciprocal pathway and the interneurons in the flexion reflex pathway (FRA) (LUNDBERG, NORRSELL and VOORHOEVE 1962). The facilitation of primary afferent depolarization (presynaptic inhibition) in Ib and II muscle and cutaneous afferents following cortical stimulation, can likewise be explained by corticospinal activation of interneurons of a spinal path from the FRA. The precise anatomical localization of these functionally different types of interneurons is as yet unknown. However, as far as can be judged from the papers of ECCLES et al. (1954; 1956; 1960), the interneurons of the Ia (AE, AI) and Ib (BE, BI) pathways appear to be located in lamina VI, while the interneurons in the FRA pathway possibly are more widely distributed in laminae IV—VI (COOMBS et al. 1956; WALL 1960). As demonstrated by NYBERG-HANSEN and BRODAL (1963), these regions of the spinal grey matter are within the area of termination of corticospinal fibres (figs. 1—2). There appears thus to be a satisfactory agreement between the anatomical and physiological observations concerning the corticospinal tract.

The recent investigations of Lundberg and collaborators (see LUNDBERG 1964b) on the interaction at the spinal level of afferent impulses from the periphery and descending impulses from the brain stem reticular formation (RF) can likewise be correlated with the present anatomical observations. Thus, following stimulation of the RF at the pontomedullary junction, presynaptic inhibition (primary afferent depolarization) of *Ia*, Ib and FRA impulses occurred by way of a descending pathway which according to LUNDBERG is situated in the ventral funiculus of the cord. Since pontine reticulospinal fibres course in this region of the cord (figs. 8A, 9) it appears likely that these fibres are involved. This assumption is strengthened by the fact that reticulospinal fibres from the pons are known to course through the region of the brain stem stimulated by the authors. However, direct reticulospinal fibres can scarcely be concerned with the *tonic* supraspinal control of spinal interneurons in the FRA pathway exerted by the brain stem RF, since impulses mediating this effect descend in the dorsolateral part of the lateral funiculus (HOLMQVIST and LUNDBERG 1959), where no reticulospinal fibres are found (figs. 8—9). Multisynaptic intersegmental pathways are probably involved in this control system.

As regards the rubrospinal and lateral vestibulospinal fibres, the physiological observations on their actions on spinal mechanisms are so far not sufficiently detailed to permit a correlation of the kind attempted above for the corticospinal and reticulospinal systems.

Several descending fibre systems are known to mediate impulses acting on the γ- as well as the α-motoneurons, *corticospinal* (Granit and Kaada 1952; Corazza et al. 1963), *reticulospinal* (Granit and Kaada 1952), *vestibulospinal* (Anderson and Gernandt 1956) and *rubrospinal* fibres (Appelberg and Kosary 1963). As emphasized by Granit (1955), there is always an intimate collaboration between the α- and γ-motoneurons in the integrative regulation of muscle tone and movements. Recently the γ-motoneurons have been found physiologically (Eccles et al. 1960a) as well as anatomically (Nyberg-Hansen 1965b) to occur intermingled with the larger α-motoneurons in the motoneuronal nuclei in lamina IX of the spinal grey matter. Together with the present anatomical findings that almost all fibres belonging to the descending supraspinal systems terminate in Rexed's laminae IV—VIII, this implies that (in the cat) the overwhelming supraspinal influences on the γ- and α-motoneurons are relayed via spinal interneurons, supposed to be localized in laminae IV—VIII. This arrangement may be assumed to permit a large degree of plasticity and freedom of the α- and γ-motoneurons and to provide a basis for a variable interplay between the α- and γ-systems, as well as between impulses converging upon the α- and γ-motoneurons from different structures of the brain.

The considerations made above are, however, subject to the following qualification: It is well known that dendrites of cells in the spinal grey matter may extend for considerable distances from the perikaryon and even beyond the confines of the lamina in which the perikaryon is situated (Cajal 1909; Lorente de Nó 1938; Aitken and Bridger 1961; Sprague 1964). Some of the terminations observed in the studies reported in this communication may, therefore, actually be on dendrites of nerve cells having their cell bodies in an adjacent lamina. For example, some terminations in lamina VII may be on dendrites of motoneurons in lamina IX extending into lamina VII. Furthermore, dendrites of motoneurons located in the ventromedial group of motoneurons situated ventrally in islands within lamina VIII, extend and ramify extensively within laminae VII—VIII (Sprague 1964). Lateral vestibulospinal and pontine reticulospinal fibres apparently ending in lamina VIII and neighbouring parts of lamina VII, may thus actually in part terminate on *dendrites* of the motoneurons. Nevertheless, there can be no doubt that a large number of terminals in a certain lamina have synaptic relations with neurons belonging to that particular lamina.

A review of the pertinent literature reveals that the terminal regions within the spinal grey matter of the various descending fibre systems appear to be more or less similar in different animal species, as far as they are known (for a more detailed discussion, see the original papers of the present author). However, anatomical (Kuypers 1960; Liu and Chambers 1964) as well as physiological (Bernhard and Bohm 1954; Preston and Whitlock 1961; Landgren, Phillips and Porter 1962) experimental studies demonstrate that in primates some corticospinal fibres terminate directly on the large α-motoneurons of the spinal cord. This seems to reflect an increasing importance of the cerebral cortex in the production of so-called "skilled voluntary movements" by the establishment of a direct route to the motoneurons, superimposed on the phylogenetically older organization of spinal segmental reflexes. At present it is not known whether the γ-motoneurons as well receive direct corticospinal impulses in primates.

e) Mode of termination of supraspinal fibres

Knowledge of the mode of termination of descending supraspinal fibres, i.e. of their synaptic relationships, must be assumed to be of value in the evaluation of neurophysiological observations, especially those obtained by the use of micro-electrodes. In the experimental studies reported here particular care has, therefore, been taken to study this subject in as great detail as possible with the methods used. Reliable information of this kind can, however, only be obtained where degenerating fibres are relatively numerous. Using the NAUTA and the GLEES techniques on transverse as well as on horizontal and sagittal sections, it has been possible to observe that the corticospinal, rubrospinal, reticulospinal, tectospinal and lateral vestibulospinal fibres form so-called pericellular arborizations and establish contact by means of terminal fibres and boutons with large as well as medium sized and small nerve cells within their laminae of termination (figs. 14—28, 30—38). While the three former fibre systems appear to form approximately equal numbers of contacts with perikarya and proximal parts of dendrites there seems to be a preponderance of axo-dendritic contacts as concerns the lateral vestibulospinal fibres (figs. 23—28). The meaning of this difference is at present not clear. It may be of functional importance. However, it should be born in mind that all close contacts between terminal structures and somata or dendrites observed in silver impregnated sections may not represent true synaptic contacts and not fulfil the electron microscopical criteria for synapses. It is, therefore, desirable to supplement silver impregnation studies on the mode of termination of degenerating fibres with corresponding electron microscopical studies. If the contacts observed in silver impregnated material represent true synapses the observations of the modes of ending made here may indicate a functional peculiarity of the lateral vestibulospinal fibres as compared to the other descending fibre systems, since recent electrophysiological observations (see ECCLES 1964) suggest that axo-somatic synapses may be inhibitory and axo-dendritic synapses exitatory, even if in the physiological sense contacts on the most proximal part of dendrites are to be considered as axo-somatic synapses.

f) The relation of the descending fibre systems to the cord to other parts of the central nervous system

Since the cell groups which give rise to fibres descending to the spinal cord are acted upon by other supraspinal structures, it is of some interest to consider how the observations made in the present studies can be correlated with other anatomical and physiological data and to consider them in a broader perspective.

1. Relations to the cerebral cortex

The red nucleus and the RF of the brain stem both receive corticofugal fibres. *Corticoreticular fibres* end in those regions of the RF which give origin to pontine and medullary reticulospinal fibres (ROSSI and BRODAL 1956; KUYPERS 1958a, b).

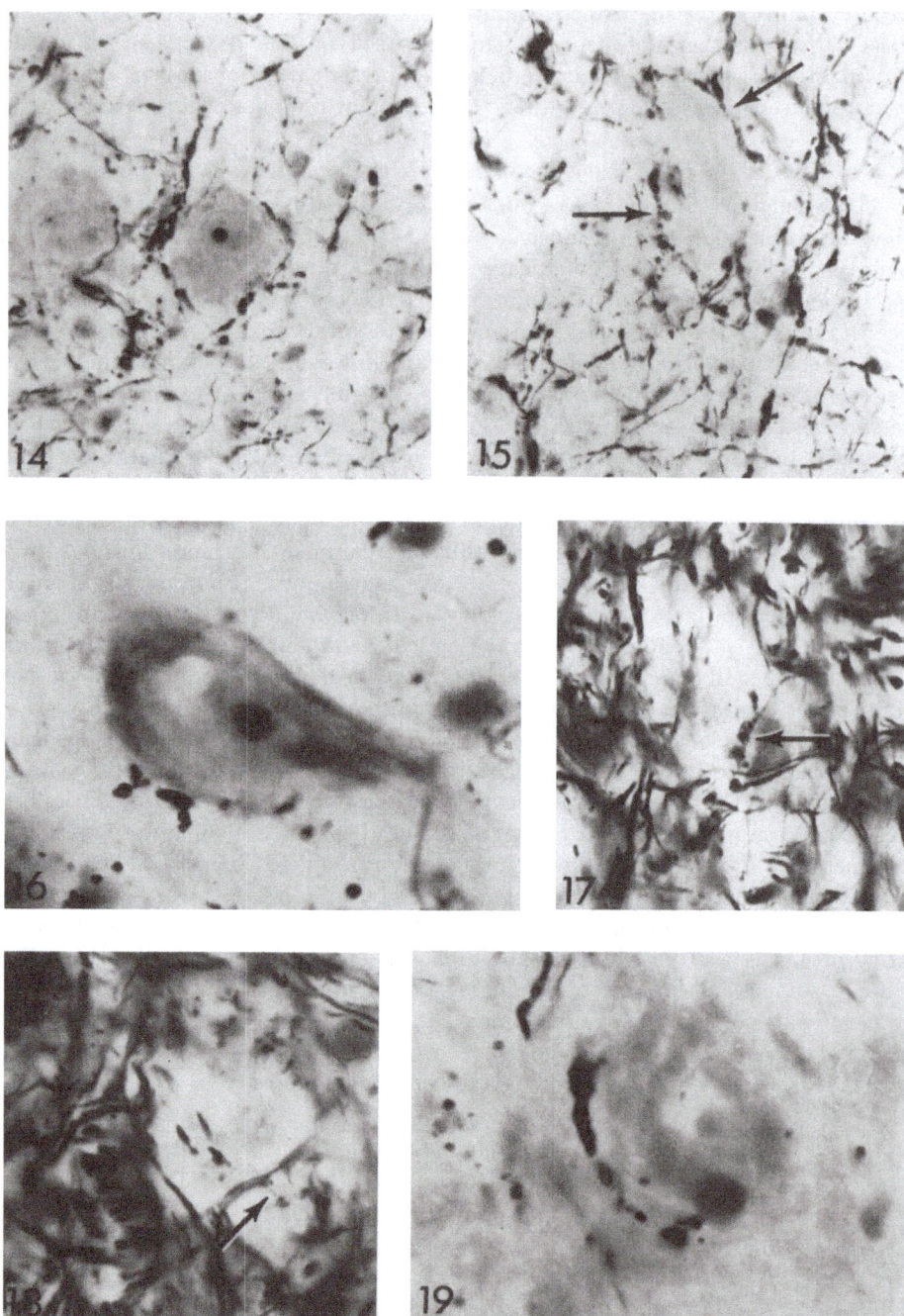

Fig. 14. Pericellular arborization around a large nerve cell laterally in lamina VI at C$_7$ of cat B.St.L. 200 with a
lesion of the primary sensorimotor cerebral cortex. NAUTA method. ×500

Fig. 15. Degenerating fibres in contact with a dendrite (arrows) of a large nerve cell laterally in lamina VI at C$_8$ of
cat B.St.L. 214 (fig. 1) with a lesion of the primary sensorimotor cerebral cortex. NAUTA mathod. ×500

Fig. 16. "Terminal degeneration" on the soma of a nerve cell in the medial half of lamina VI at C$_8$ of cat B.St.L. 200
(cp. fig. 14). NAUTA-GYGAX method. ×1200 (oil immersion)

The corticorubral projection has recently been shown to be arranged in a somato-topical pattern (RINVIK and WALBERG 1963), which fits in with the somatotopical organization of the rubrospinal projection demonstrated earlier (POMPEIANO and BRODAL 1957a). As described in the present paper corticospinal fibres from the "motor" area of the primary sensorimotor cortex, medullary reticulospinal and rubrospinal fibres have partly the same areas of termination within the spinal grey matter (laminae V—VII, see fig. 12B). As mentioned above, this region of the grey matter appears to harbour interneurons intercalated in pathways excitatory to flexor and inhibitory to extensor motoneurons. It is interesting to note that according to physiological observations these three fibre systems all act on the spinal cord by facilitating its flexor mechanisms, while the pontine reticulo-spinal and lateral vestibulospinal fibres facilitate the extensor mechanisms. Thus, the cerebral cortex may influence spinal extensor mechanisms by way of cortico-pontine-reticulospinal fibres, while the spinal flexor mechanisms are influenced by corticomedullary-reticulospinal and cortico-rubrospinal fibres as well as by the direct corticospinal projection. The two cortico-reticulospinal projections lack a somatotopical organization, while the two latter pathways betray such locali-zation. The possible role of the somatotopically organized cortico-rubrospinal projection in relation to the preserved capacity of performing "skilled voluntary movements" following pedunculotomies in man and monkey (BUCY and KEP-LINGER 1961; BUCY, KEPLINGER and SIQUERA 1964) has been discussed elsewhere (BRODAL 1963b; RINVIK and WALBERG 1963). Both projections presumably collaborate intimately in mediating cortical influences on spinal flexor mechanisms. The similar sites of termination in the spinal grey matter of these two fibre systems fit in with and lend support to this view (fig. 13).

2. Relations to the cerebellum

The principal dissimilarities with regard to the sites of termination in the spinal grey matter of the *rubrospinal* (fig. 4) and the *lateral vestibulospinal* fibres (fig. 5, see also fig. 12A and B) may be assumed to reflect basic functional features. This assumption is strengthened when the role of the red nucleus and the lateral vestibular nucleus as relay stations for *cerebello-spinal* impulses controlling flexor and extensor motoneurons, respectively, is taken into account.

The *lateral vestibulospinal fibres* are the final common path in two routes from the cerebellar vermis to the lateral vestibular nucleus, one direct, the other relayed in the fastigial nucleus, while the *rubrospinal fibres* are the final link in the follow-ing pathway: intermediate cerebellar cortex — interpositus nucleus — red nucleus. The cerebello-rubrospinal as well as the cerebello-vestibulospinal projections are somatotopically organized throughout, and are thus well suited to convey impulses for localized cerebellar control of spinal *flexor* and *extensor motoneurons*, respec-tively (see MAFFEI and POMPEIANO 1962, and BRODAL, POMPEIANO and WALBERG

Fig. 17. "Terminal" degenerating fibre (arrow) between two nerve cells in lamina V at L$_5$ of cat B.St.L. 200 (cp. fig. 14). GLEES method. $\times$ 1200 (oil immersion)

Fig. 18. "Terminal degeneration" on the soma of a nerve cell in lamina VI at L$_5$ of cat B.St.L. 200 (cp. fig. 14). Arrow points to the tail of a degenerating bouton (left). GLEES method. $\times$ 1200 (oil immersion)

Fig. 19. "Preterminal degeneration" on a small nerve cell laterally in lamina VI at C$_7$ of cat B.St.L. 268 with a lesion of the red nucleus. NAUTA method. $\times$ 1200 (oil immersion)

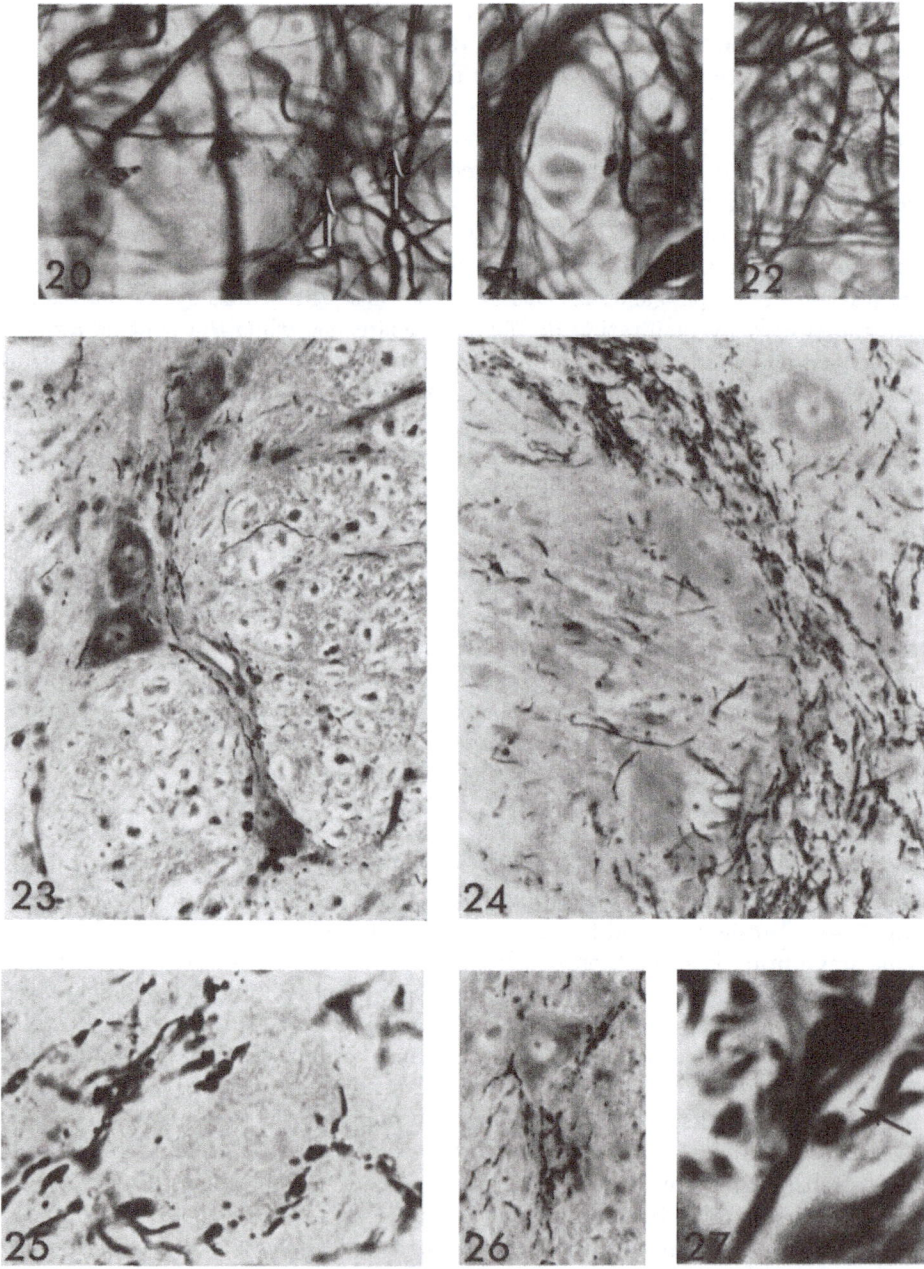

Fig. 20. Fine degenerating fibre (arrows) and (to the left) a degenerating fragment, presumably a bouton, medially in lamina VI in the cervical cord of cat B.St.L. 126 with a lesion of the red nucleus. GLEES method. ×1100 (oil immersion)

Fig. 21. Degenerating bouton in contact with a small nerve cell medially in lamina VI of cat B.St.L. 120 with a lesion of the red nucleus. GLEES method. ×1100 (oil immersion)

Fig. 22. Degenerating fragments, presumably boutons, medially in lamina VI in the cervical cord of cat B.St.L. 126 (cp. fig. 20). GLEES method. ×1100 (oil immersion)

1962, respectively)[1]. The dissimilar sites of termination of rubrospinal (laminae V —VII) and lateral vestibulospinal fibres (laminae VII—VIII), demonstrated in the present anatomical investigations thus are in complete accord with the physiological data on the reciprocal control of spinal motoneurons exerted by these two descending fibre systems[2].

g) Central control of sensory impulses

Stimulations of the *sensorimotor area* of the cerebral cortex and the *RF* of the brain stem for some time have been known to *depress* the transmission of afferent impulses in ascending fibre tracts of the spinal cord (see HAGBARTH 1960). According to recent investigations this supraspinal control of sensory impulses is exerted particularly on ascending spinal pathways influenced by flexion reflex afferents (FRA) (see LUNDBERG 1964b). Transmission of FRA impulses to all ascending tracts has been found to be tonically inhibited by a multisynaptic intersegmental control system which descends in the dorsolateral part of the lateral funiculus considered on p. 31 (HOLMQVIST, LUNDBERG and OSCARSSON 1960a). Furthermore, the spinal interneurons in the FRA pathways mediating effects to the motoneurons and to the ascending tracts are influenced from the periphery and from supraspinal structures in the same way. Thus, corticospinal impulses facilitate transmission of FRA impulses to the spinobulbar (reticular) tract ascending in the ventral funiculus and to neurons of the dorsal spinocerebellar tract influenced by FRA (LUNDBERG, NORRSELL and VOORHOEVE 1963), while they inhibit transmission of FRA impulses to the ventral spinocerebellar tract (MAGNI and OSCARSSON 1961; LUNDBERG, NORRSELL and VOORHOEVE 1963).

---

[1] The lack of a somatotopical pattern within the cerebelloreticular projection to the nucleus reticularis gigantocellularis of the medulla oblongata (WALBERG, POMPEIANO, WESTRUM and HAUGLIE-HANSSEN 1962) and within the reticulospinal projection (TORVIK and BRODAL 1957) makes it likely that the cerebello-reticulospinal fibre systems contribute to a more general background activity of the spinal mechanisms upon which impulses in the somatotopically organized pathways may play.

[2] Although no information is available on this point, one may speculate upon whether the rubrospinal and lateral vestibulospinal fibres from large and small neurons in their nuclei of origin have identical sites and mode of termination within the spinal grey matter, and whether large and small neurons in the lateral vestibular and red nucleus are related to the activation of α- and γ-motoneurons, respectively. The question may, furthermore, be raised whether this has something to do with the role of the cerebellum in integrating and controlling the activity of the α- and γ-motoneurons (GRANIT, HOLMGREN and MERTON 1955).

Fig. 23. Degenerating fibres entering the spinal grey matter (to the left) along the dendrites of nerve cells medially in lamina VIII in a transverse section at L_7 of cat B.St.L. 254 with a lesion of the lateral vestibular nucleus. NAUTA method. ×300

Fig. 24. Degenerating fibres acquiring a longitudinal course after entering the spinal grey matter medially in lamina VIII at L_6 of cat B.St.L. 307 (fig. 5) with a lesion of the lateral vestibular nucleus. Sagittal section. NAUTA method. ×300

Fig. 25. Pericellular arborization around the perikaryon of a nerve cell in lamina VIII at L_6 of cat B.St.L. 307 (cp. fig. 24). NAUTA method. ×1400 (oil immersion)

Fig. 26. "Preterminal degeneration" on the perikaryon and especially along a proximal dendrite of a small nerve cell in lamina VII at L_6 of cat B.St.L. 254 (cp. fig. 23). NAUTA method. ×300

Fig. 27. Degenerating bouton with a fragmented tail (arrow) on a dendrite of a nerve cell in lamina VIII at C_7 of cat B.St.L. 287 with a lesion of the lateral vestibular nucleus. GLEES method. ×1200 (oil immersion)

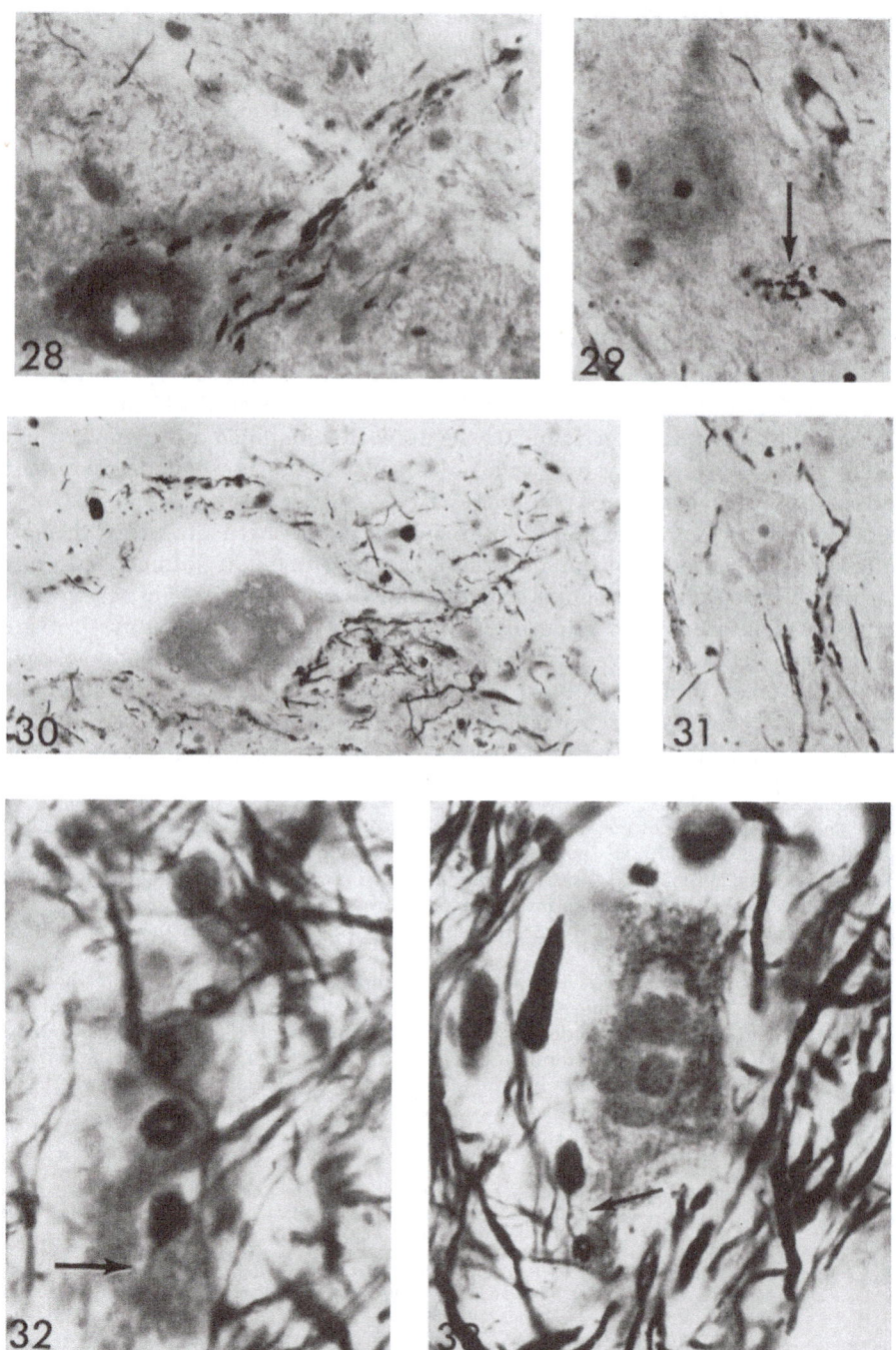

Fig. 28. Pericellular arborization along the proximal part of a dendrite of a nerve cell in lamina VIII at L₇ of cat B.St.L. 287 (cp. fig. 27). NAUTA method. ×570

Fig. 29. Degeneration on a proximal dendrite (arrow) of a nerve cell in lamina VIII at C₇ of cat B.St.L. 311 (fig. 7) with a lesion of the medial vestibular nucleus. NAUTA method. ×600

Cortico-rubrospinal impulses, however, facilitate the latter transmission (MAGNI and OSCARSSON 1961). Postsynaptic inhibition in interneurons appears to be involved (LUNDBERG 1964a). The interneurons are believed to be intercalated in spinal flexion reflex pathways (LUNDBERG 1964a; MAGNI and OSCARSSON 1961), and appear to be localized in laminae IV—VI, where NYBERG-HANSEN and BRODAL (1963, 1964) found corticospinal and rubrospinal fibres to terminate (figs. 1, 2 and 4). There appears thus to be satisfactory agreement between the anatomical and physiological observations that the corticospinal and rubrospinal fibres participate in the supraspinal control of sensory synaptic relays in the spinal cord[1].

Concerning the ventral spinobulbar (reticular) tract polysynaptically influenced by FRA impulses (LUNDBERG and OSCARSSON 1962), NYBERG-HANSEN (1965a) put forward evidence that the descending control system in the ventral funiculus which monosynaptically activates the neurons of the tract (HOLMQVIST, LUNDBERG and OSCARSSON 1960b), probably is the pontine reticulospinal tract. This assumption is compatible with the physiological observations presented by LUNDBERG and his collaborators.

As considered in a preceding section (Descending supraspinal impulses) pontine reticulospinal fibres are probably involved in *presynaptic inhibition* of 1a, 1b and FRA impulses in the cord. Furthermore, the recent findings of ANDERSEN, ECCLES and SEARS (1964) that presynaptic inhibition of Ib, II and III afferent impulses occurs in the dorsal horn of the cord following stimulation of the *"sensory"* area of the primary sensorimotor cortex, may be suggested to have its morphological correlate in the observation of NYBERG-HANSEN and BRODAL (1963) that corticospinal fibres from the "sensory" area terminate chiefly in laminae IV—V (while those from the "motor" area in general terminate more ventrolaterally in laminae V—VII, fig. 2).

h) "Direction specific movements"

Tectospinal, interstitiospinal and medial vestibulospinal fibres betray some common features. They descend medially in the ventral funiculus and terminate mainly medially in the spinal grey matter. These fibre systems are probably

---

[1] Unfortunately, little is known as concerns the exact sites of origin within the spinal grey matter of secondary sensory neurons of the various ascending tracts of the spinal cord (except the dorsal spinocerebellar tract). Usually the perikarya are assumed to be situated in the dorsal horn (laminae I—IV). Recently, however, HUBBARD and OSCARSSON (1961) report physiological evidence for ventral spinocerebellar tract neurons located laterally in laminae V—VII (see also OSCARSSON 1964). Exact anatomical data on the localization of spinal neurons sending their axons rostrally to the brain stem are desirable before a closer correlation with the physiological observations can be attempted.

Fig. 30. Pericellular arborization close to the perikaryon and along a proximal dendrite of a nerve cell medially in lamina VII at C_7 of cat B.St.L. 299 (fig. 8a) with a lesion of the nucleus reticularis pontis caudalis. NAUTA method. ×400

Fig. 31. Pericellular arborization along a proximal dendrite of a small nerve cell in lamina VIII at L_7 of cat B.St.L. 322 with a lesion of the nucleus reticularis pontis caudalis. NAUTA method. ×480

Fig. 32. Degenerating bouton with a fragmented tail (arrow) on the soma of a small nerve cell in lamina VIII at C_8 of cat B.St.L. 338 with a lesion of the nucleus reticularis pontis caudalis. GLEES method. ×1550 (oil immersion)

Fig. 33. Degenerating irregular solid bouton with an irregular tail (arrow) on the soma of a small nerve cell centrally in lamina VII at C_8 of cat B.St.L. 318 (fig. 8b) with a lesion of the nucleus reticularis gigantocellularis. GLEES method. ×1550 (oil immersion)

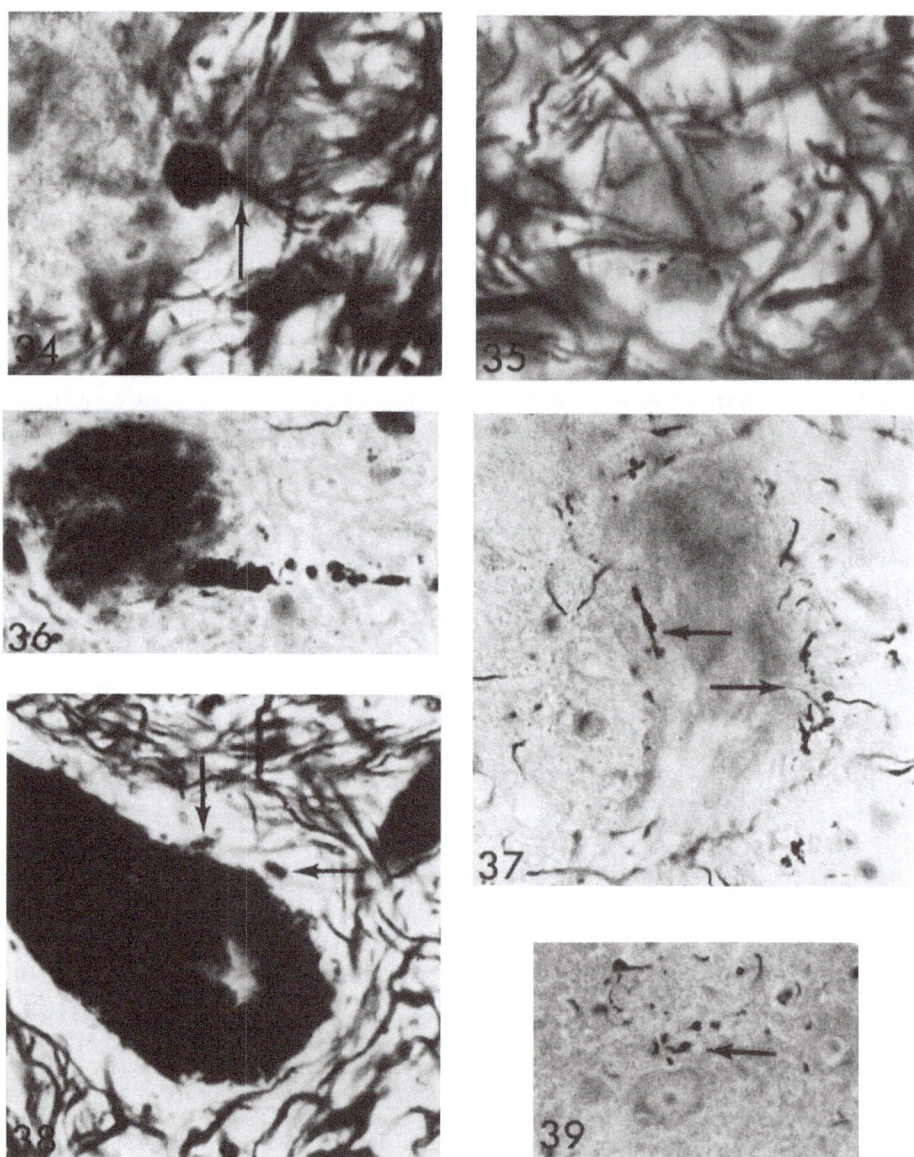

Fig. 34. Degenerating irregular solid bouton with an irregular tail (arrow) on the soma of a large nerve cell centrally in lamina VII at C_8 of cat B.St.L. 320 with a lesion of the nucleus reticularis gigantocellularis. GLEES method. × 1550 (oil immersion)

Fig. 35. Terminal degenerating fibres on a nerve cell in lamina VIII at C_8 of cat B.St.L. 338 (cp. fig. 32). GLEES method. × 1550 (oil immersion)

Fig. 36. "Terminal degeneration" on the soma of a nerve cell centrally in lamina VII at C_5 of cat B.St.L. 293 with a lesion of the superior colliculus. NAUTA method. × 950

Fig. 37. "Preterminal degeneration" on a dendrite (arrows) of a nerve cell laterally in lamina VII at C_3 of cat B.St.L. 276 with a lesion of the superior colliculus. NAUTA method. × 600

Fig. 38. Degenerating boutons (arrows) with their degenerating tails slightly out of focus, on a large nerve cell centrally in lamina VII at C_5 of cat B.St.L. 293 (cp. fig. 36). GLEES method. × 1450 (oil immersion)

Fig. 39. "Preterminal degeneration" (arrow) in the neighbourhood of nerve cells belonging to the intermediomedial nucleus at Th_5 of cat B.St.L. 316 (fig. 11) with a lesion of the interstitial nucleus of CAJAL. NAUTA method. × 600

concerned with so-called "direction specific movements". Thus space-orientating movements of the eyes, head and body may be elicited by stimulation of the superior colliculus (APTER 1946), interstitial nucleus (HASSLER and HESS 1954) and the medial vestibular nucleus (MONTANDON and MONNIER 1964), and are reflexely evoked in the normal animal by afferent impulses from the retina and the labyrinth[1]. The descending impulses to the cervical motoneurons engaged in the head movements and the motoneurons of the cord innervating the axial muscles concerned with rotation around the longitudinal axis of the body, are probably conveyed by tectospinal, medial vestibulospinal and interstitiospinal fibres terminating on interneurons mainly in laminae VII—VIII (figs. 10, 7, 11). The latter fibre system descends to sacral levels and may thus act on all levels of the cord in contrast to the two former systems which mainly act on the cervical motoneurons responsible for adjustments of the position of the head.

V. Summary and conclusions

A review is given of the anatomy of descending supraspinal fibre systems to the spinal cord in the cat, based on experimental investigations of the author and his collaborators. The origin, course and sites and mode of termination have been studied by the use of silver impregnation technique following lesions of various supraspinal structures. Particular attention has been devoted to a precise mapping of the sites of ending with reference to REXED's architectonic map of the cord (REXED 1952, 1954).

Corticospinal fibres from the primary sensorimotor cortex descend to the lowest segments of the lumbosacral enlargement and terminate in laminae IV—VII. Fibres from the "motor" cortex terminate in laminae V—VII, mainly laterally, those from the "sensory" cortex in laminae IV—VI, chiefly medially.

Rubrospinal fibres descend contralaterally to low lumbosacral levels and terminate in the lateral parts of laminae V—VI and centrally and laterally in lamina VII. The observation of POMPEIANO and BRODAL (1957a) that the rubrospinal projection is somatotopically organized is confirmed.

Lateral vestibulospinal fibres originate in the *lateral* vestibular nucleus and descend in the homolateral ventral funiculus to sacral levels. The fibres terminate in the entire lamina VIII and the neighbouring parts of lamina VII. The lateral vestibulospinal fibre system is organized in a somatotopical manner, as first shown by POMPEIANO and BRODAL (1957b).

Medial vestibulospinal fibres take origin from the *medial* vestibular nucleus and descend bilaterally in the medial longitudinal fasciculus in the ventral funiculus to midthoracic levels. The fibres terminate in dorsal parts of lamina VIII and the adjacent part of lamina VII.

Reticulospinal fibres are divided into two contingents: *Pontine fibres*, chiefly from the nucleus reticularis pontis caudalis, descend to sacral levels in the ventral funiculus, mainly homolaterally. Some cross to the contralateral grey matter by way of the anterior commissure. The fibres terminate in the entire lamina VIII and adjacent parts of lamina VII. *Medullary reticulospinal fibres* originate chiefly in the nucleus reticularis gigantocellularis of the medulla and descend bilaterally

---

[1] These movements may be part of a general orienting behaviour.

in the ventral half of the lateral funiculus to sacral levels. The fibres terminate chiefly in lamina VII. A few terminations are found in lamina IX. No reticulospinal fibres originate in the RF of the mesencephalon.

Tectospinal fibres come from the superior colliculus and descend ventrally in the contralateral ventral funiculus to the lowest segments of the cervical enlargement. Most fibres terminate in the four upper cervical segments in lamina VII and the neighbouring parts of lamina VI and VIII.

Interstitiospinal fibres take origin from the interstitial nucleus of Cajal in the rostral mesencephalon, descend bilaterally in the dorsal third of the ventral funiculus to sacral levels, and terminate in dorsal parts of lamina VIII and the adjacent part of lamina VII.

No fibres from any of the systems studied end in the large lateral lamina IX except for an extremely small number of medullary reticulospinal fibres which possibly terminate on α-motoneurons. Fibres from all the systems studied appear to terminate on neurons of all sizes within their laminae of termination. Both axo-somatic contacts and contacts with proximal dendrites are found. No descending fibres terminate on neurons belonging to the column of Clarke or the intermediolateral cell column.

The findings are discussed from a functional point of view, and attempts are made to correlate the observations with other data. Among other things, attention is drawn to the fact that phylogenetically old systems as the vestibulospinal and pontine reticulospinal fibres, descend in the ventral funiculus and terminate ventromedially, in lamina VIII and neighbouring parts of lamina VII. These fibre systems convey impulses which are chiefly facilitatory to extensor and inhibitory to flexor motoneurons. Phylogenetically younger pathways, such as rubrospinal and corticospinal fibres, mediate mainly the opposite effects, descend in the lateral funiculus and terminate dorsolaterally in the spinal grey matter corresponding to laminae IV—VII. These anatomical and physiological similarities reflect basic functional features in the intrinsic organization of the spinal cord, especially as concerns a reciprocal organization of motor performances.

A broader perspective of these and other aspects of the cord's organization is obtained when the present anatomical findings are considered in relation to some physiological data on the supraspinal control of spinal flexor and extensor mechanisms. The relations of the red nucleus, lateral vestibular nucleus and the brain stem RF to the cerebral cortex and the cerebellum are discussed from this point of view.

The tectospinal, medial vestibulospinal and interstitiospinal fibres are assumed to be related particularly to so-called "direction specific movements".

The role of the cerebral cortex, the red nucleus and the brain stem reticular formation in controlling sensory impulses, and the interaction at the spinal level of afferent impulses from the periphery and descending impulses from supraspinal structures, are briefly commented upon and correlated with the present anatomical findings.

The anatomical observations on the supraspinal descending fibre systems presented in this review are in general agreement with conclusions reached in recent neurophysiological investigations.

Acknowledgement. The original investigations on which the present publication is based, have been carried out at the Anatomical Institute, University of Oslo. I have been very fortunate in having the opportunity to work under the guidance of Professor ALF BRODAL, M. D. His never failing interest, encouragement and criticism have been inspiring and of great value to me throughout the progress of my studies.

I also wish to express my gratitude to Professor JAN JANSEN, M. D., Head of the Anatomical Institute, for good working facilities. Thanks are also due to members of the staff for valuable discussions.

The expert assistance given me by the technical staff of the Institute, especially Miss KARI HVERVEN who has made most of the sections, Miss ODDLAUG GORSET in the typewriting, Mrs. INGER EGGUM in making the drawings and Mr. EINAR RISNES who has done the photographic work, is greatly appreciated.

My work in the Anatomical Institute has been carried out under a special research education grant from the Norwegian Research Council for Science and the Humanities.

Bibliography

AITKEN, J. T., and J. E. BRIDGER: Neuron size and neuron population density in the lumbosacral region of the cat's spinal cord. J. Anat. (Lond.) **95**, 38—53 (1961).

ALTMAN, J., and M. B. CARPENTER: Fiber projections of the superior colliculus in the cat. J. comp. Neurol. **116**, 157—177 (1961).

ANDERSEN, P., J. C. ECCLES, and T. A. SEARS: Cortically evoked depolarization of primary afferent fibers in the spinal cord. J. Neurophysiol. **27**, 63—77 (1964).

ANDERSON, S., and B. E. GERNANDT: Ventral root discharge in response to vestibular and proprioceptive stimulation. J. Neurophysiol. **19**, 524—543 (1956).

APPELBERG, B., and I. Z. KOSARY: Excitation of flexor fusimotor neurons by electrical stimulation in the red nucleus. Acta physiol. scand. **59**, 445—453 (1963).

APTER, J. T.: Eye movements following strychninization of the superior colliculus of cat. J. Neurophysiol. **9**, 73—86 (1946).

BATES, J. A. V.: Observations on the excitable cortex in man. In: Lectures on the scientific basis of medicine, vol. 5, p. 333—347. London: Athlone Press 1957.

BERNHARD, C. G., and E. BOHM: Cortical representation and functional significance of the corticomotoneuronal system. Arch. Neurol. Psychiat. (Chic.) **72**, 473—502 (1954).

—, and B. REXED: The localization of the premotor interneurons discharging through the peroneal nerve. J. Neurophysiol. **8**, 387—392 (1945).

BRODAL, A.: Modification of Gudden method for study of cerebral localization. Arch. Neurol. Psychiat. (Chic.) **43**, 46—58 (1940).

— The reticular formation of the brain stem. Anatomical aspects and functional correlation. The Henderson trust lectures. Edinburgh and London: Oliver & Boyd 1957.

— Anatomical observations on the vestibular nuclei, with special reference to their relations to the spinal cord and the cerebellum. Acta oto-laryng. (Stockh.), Suppl. **192**, 24—51 (1963 a).

— Er pyramidebanen nødvendig for utførelsen av finere vilkårlige bevegelser? Nord. Med. **70**, 1342 (1963 b).

— Anatomical organization and fiber connections of the vestibular nuclei. In: Neurological aspects of auditory and vestibular disorders. Eds. W. S. FIELDS and B. R. ALFORD. Springfield (Ill.): Ch. C. Thomas 1964.

—, and A. CHR. GOGSTAD: Rubrocerebellar connections. An experimental study in the cat. Anat. Rec. **118**, 455—486 (1954).

—, and O. POMPEIANO: The vestibular nuclei in the cat. J. Anat. (Lond.) **91**, 438—454 (1957).

— — and F. WALBERG: The vestibular nuclei and their connections, anatomy and functional correlations. The Henderson trust lectures. Edinburgh and London: Oliver & Boyd 1962.

BUCHANAN, B. R.: The course of the secondary vestibular fibers in the cat. J. comp. Neurol. **67**, 183—204 (1937).

BUCY, P. C., and J. KEPLINGER: Section of the cerebral peduncles. Arch. Neurol. (Chic.) 5, 132—139 (1961).
— — and E. B. SIGUEIRA: Destruction of the "pyramidal" tract in man. J. Neurosurg. 21, 385—398 (1964).
CAJAL, S. R.: Histologie du système nerveux de l'homme et des vertébrés. Paris: Maloine 1909.
CHAMBERS, W. W., and C.-N. LIU: Corticospinal tract of the cat. An attempt to correlate the pattern of degeneration with deficits in reflex activity following neocortical lesions. J. comp. Neurol. 108, 23—55 (1957).
COOMBS, J. S., D. A. CURTIS, and S. LANDGREN: Spinal cord potentials generated by impulses in muscle and cutaneous afferent fibers. J. Neurophysiol. 19, 452—467 (1956).
CORAZZA, R., E. FADIGA, and P. L. PARMEGGIANI: Patterns of pyramidal activation of cat's motoneurons. Arch. ital. Biol. 101, 337—364 (1963).
ECCLES, J. C.: The physiology of nerve cells. Baltimore: Johns Hopkins Press 1957.
— The physiology of synapses. Berlin-Göttingen-Heidelberg: Springer 1964.
— R. M. ECCLES, A. IGGO, and A. LUNDBERG: Electrophysiological studies on gamma moto-neurons. Acta physiol. scand. 50, 32—40 (1960a).
— — and A. LUNDBERG: Types of neurone in and around the intermediate nucleus of the lumbosacral cord. J. Physiol. (Lond.) 154, 89—114 (1960b).
— P. FATT, and S. LANDGREN: Central pathway for direct inhibitory action of impulses in largest afferent nerve fibres to muscle. J. Neurophysiol. 19, 75—98 (1956).
— — — and G. J. WINSBURY: Spinal cord potentials generated in the large muscle afferents. J. Physiol. (Lond.) 125, 590—606 (1954).
FERRARO, A., B. L. PACELLA, and S. E. BARRERA: Effects of lesions of the medial vestibular nucleus. An anatomical and physiological study in Macacus Rhesus monkeys. J. comp. Neurol. 73, 7—36 (1940).
GLEES, P.: Terminal degeneration within the central nervous system as studied by a new silver method. J. Neuropath. exp. Neurol. 5, 54—59 (1946).
GRANIT, R.: Receptors and sensory perception. New Haven: Yale University Press 1955.
— B. HOLMGREN, and P. A. MERTON: The two routes for excitation of muscle and their sub-servience to the cerebellum. J. Physiol. (Lond.) 130, 213—224 (1955).
—, and B. R. KAADA: Influence of stimulation of central nervous structures on muscle spindles in cat. Acta physiol. scand. 27, 130—160 (1952).
HAGBARTH, K. E.: Centrifugal mechanisms of sensory control. Ergebn. Biol. 22, 47—66 (1960).
HASSLER, R., u. W. R. HESS: Experimentelle und anatomische Befunde über die Dreh-bewegungen und ihre nervösen Apparate. Arch. Psychiat. Z. Neurol. 192, 488—526 (1954).
HINMAN, A., and M. B. CARPENTER: Efferent fiber projections of the red nucleus in the cat. J. comp. Neurol. 113, 61—82 (1959).
HOLMQVIST, B., and A. LUNDBERG: On the organization of the supraspinal inhibitory control of interneurones of various spinal reflex arcs. Arch. ital. Biol. 97, 340—356 (1959).
— — and O. OSCARSSON: Supraspinal inhibitory control of transmission to three ascending spinal pathways influenced by the flexion reflex afferents. Arch. ital. Biol. 98, 60—80 (1960a).
— — — A supraspinal control system monosynaptically connected with an ascending spinal pathway. Arch. ital. Biol. 98, 402—422 (1960b).
HUBBARD, J. I., and O. OSCARSSON: Localization of the cell bodies of the ventral spino-cerebellar tract in lumbar segments of the cat. J. comp. Neurol. 118, 199—204 (1962).
KAPPERS, C. U. ARIËNS, G. C. HUBER, and E. C. CROSBY: The comparative anatomy of the nervous system of vertebrates, including man. New York: Macmillan Co. 1936.
KOHNSTAMM, O.: Ueber die Coordinationskerne des Hirnstammes und die absteigenden Spinalbahnen. Mschr. Psychiat. Neurol. 8, 261—293 (1900).
KUYPERS, H. G. J. M.: An anatomical analysis of corticobulbar connexions to the pons and the lower brain stem in the cat. J. Anat. (Lond.) 92, 198—218 (1958a).
— Corticobulbar connexions from the pericentral cortex to the pons and lower brain stem in monkey and chimpanzee. J. comp. Neurol. 110, 221—256 (1958b).
— Central cortical projections to motor and somatosensory cell groups. Brain 83, 161—184 (1960).

KUYPERS, H. G. J. M., W. R. FLEMING, and J. W. FARINHOLT: Subcortical projections in the Rhesus monkey. J. comp. Neurol. 118, 107—137 (1962).

LANDGREN, S., C. G. PHILLIPS, and R. PORTER: Minimal synaptic actions of pyramidal impulses on some alpha motoneurones of the baboon's hand and forearm. J. Physiol. (Lond.) 161, 91—111 (1962).

LEWANDOWSKY, M.: Untersuchungen über die Leitungsbahnen des Truncus cerebri und ihren Zusammenhang mit denen der Medulla spinalis und des Cortex cerebri. Denkschr. med.-naturw. Ges. Jena, Neurobiol. Arb., Ser. II 1, 63—150 (1904).

LIU, C. N., and W. W. CHAMBERS: An experimental study of the corticospinal system in the monkey (Macaca mulatta). J. comp. Neurol. 123, 257—284 (1964).

LORENTE DE NÓ, R.: Synaptic stimulation of motoneurons as a local process. J. Neurophysiol. 1, 195—206 (1938).

LUNDBERG, A.: Ascending spinal hindlimb pathways in the cat. Progr. Brain Res. 12, 135—163 (1964a).

— Supraspinal control of transmission in reflex paths to motoneurones and primary afferents. Progr. Brain Res. 12, 197—219 (1964b).

— U. NORSELL, and P. VOORHOEVE: Pyramidal effects on lumbo-sacral interneurons activated by somatic afferents. Acta physiol. scand. 56, 220—229 (1962).

— — — Effects from the sensorimotor cortex on ascending spinal pathways. Acta physiol. scand. 59, 462—473 (1963).

—, and O. OSCARSSON: Two ascending spinal pathways in the ventral part of the cord. Acta physiol. scand. 54, 270—286 (1962).

—, and P. VOORHOEVE: Effects from the pyramidal tract on spinal reflex arcs. Acta physiol. scand. 56, 201—219 (1962).

MAFFEI, L., and O. POMPEIANO: Cerebellar control of flexor motoneurons. An analysis of the postural responses to stimulation of the paramedian lobule in the decerebrate cat. Arch. ital. Biol. 100, 476—509 (1962).

MAGNI, F., and O. OSCARSSON: Cerebral control of transmission to the ventral spino-cerebellar tract. Arch. ital. Biol. 99, 369—396 (1961).

MONTANDON, P., and M. MONNIER: Correlation of the diencephalic nystamogenic area with the bulbovestibular nystagmogenic area. Brain 87, 673—690 (1964).

NAUTA, W. J. H.: Silver impregnation of degenerating axons. In: New research techniques of neuroanatomy, p. 17—26. Ed. WILLIAM F. WINDLE. Springfield (Ill.): Ch. C. Thomas 1957.

—, and P. A. GYGAX: Silver impregnation of degenerating axons in the central nervous system: a modified technique. Stain Technol. 29, 91—93 (1954).

NYBERG-HANSEN, R.: Origin and termination of fibers from the vestibular nuclei descending in the medial longitudinal fasciculus. An experimental study with silver impregnation methods in the cat. J. comp. Neurol. 122, 355—357 (1964a).

— The location and termination of tectospinal fibers in the cat. Exp. Neurol. 9, 212—227 (1964b).

— Sites and mode of termination of reticulo-spinal fibers in the cat. An experimental study with silver impregnation methods. J. comp. Neurol. 124, 71—100 (1965a).

— Anatomical demonstration of gamma motoneurones in the cat's spinal cord. Exp. Neurol. 13, 71—81 (1965b).

— Sites of termination of interstitiospinal fibers in the cat. An experimental study with silver impregnation methods. Arch. ital. Biol. (1966) (in press).

—, and A. BRODAL: Sites of termination of corticospinal fibers in the cat. An experimental study with silver impregnation methods. J. comp. Neurol. 120, 369—391 (1963).

— — Sites and mode of termination of rubrospinal fibres in the cat. An experimental study with silver impregnation methods. J. Anat. (Lond.) 98, 235—253 (1964).

—, and T. A. MASCITTI: Sites and mode of termination of fibers of the vestibulospinal tract in the cat. An experimental study with silver impregnation methods. J. comp. Neurol. 122, 369—387 (1964).

—, and E. RINVIK: Some comments on the pyramidal tract, with special reference to its individual variations in man. Acta neurol. scand. 39, 1—30 (1963).

OSCARSSON, O.: Differential course and organization of uncrossed and crossed long ascending spinal tracts. Progr. Brain Res. 12, 164—176 (1964).

Pearce, G. W., and P. Glees: The termination of the crossed tecto-spinal tract in the spinal cord of the cat. (Abstract.) J. Anat. (Lond.) 90, 565—566 (1956).

Pompeiano, O.: Analisi degli effetti della stimolazione elettrica del nucleo rosso nel gatto decerebrato. R. C. Accad. naz. Lincei, Cl. Sci. fis., mat. nat., Ser. VIII 22, 100—103 (1957).

—, and A. Brodal: Experimental demonstration of a somatotopical origin of rubrospinal fibers in the cat. J. comp. Neurol. 108, 225—252 (1957a).

— — The origin of vestibulospinal fibres in the cat. An experimental-anatomical study, with comments on the descending medial longitudinal fasciculus. Arch. ital. Biol. 95, 166—195 (1957b).

Preston, J. B., and D. G. Whitloch: Intracellular potentials recorded from motoneurons following precentral gyrus stimulation in primate. J. Neurophysiol. 24, 91—100 (1961).

— — A comparison of motor cortex effects on slow and fast muscle innervations in the monkey. Exp. Neurol. 7, 327—341 (1963).

Probst, M.: Experimentelle Untersuchungen über die Schleifenendigung, die Hauben-bahnen, das dorsale Längsbündel und die hintere Commissur. Arch. Psychiat. Nervenkr. 33, 1—57 (1900).

Rasdolsky, J.: Über die Endigung der extraspinalen Bewegungssysteme im Rückenmark. Z. ges. Neurol. Psychiat. 86, 360—374 (1923).

Rasmussen, A. T.: Tractus tecto-spinalis in the cat. J. comp. Neurol. 63, 501—526 (1936).

Rexed, B.: The cytoarchitectonic organization of the spinal cord in the cat. J. comp. Neurol. 96, 415—496 (1952).

— A cytoarchitectonic atlas of the spinal cord in the cat. J. comp. Neurol. 100, 297—380 (1954).

Rinvik, E., and F. Walberg: Demonstration of a somatotopically arranged corticorubral projection in the cat. An experimental study with silver methods. J. comp. Neurol. 120, 393—407 (1963).

Rossi, G. F., and A. Brodal: Corticofugal fibres to the brain-stem reticular formation. An experimental study in the cat. J. Anat. (Lond.) 90, 42—62 (1956).

—, and A. Zanchetti: The brain stem reticular formation. Anatomy and physiology. Arch. ital. Biol. 95, 199—435 (1957).

Sasaki, K., A. Namikawa, and S. Hashiramoto: The effect of midbrain stimulation upon alpha motoneurons in lumbar spinal cord of the cat. Jap. J. Physiol. 10, 303—316 (1960).

Schimert, J. S.: Die Endigung des Tractus vestibulospinalis. Z. Anat. Entwickl.-Gesch. 108, 761—767 (1938).

Schueren, A. van der: Étude anatomique du faisceau longitudinal posterieur. Névraxe 13, 183—309 (1912).

Sherrington, C. S.: The intergrative action of the nervous system. Silliman memorial lectures. New Haven: Yale University Press; London: Constable 1906.

Sprague, J. M.: The terminal fields of dorsal root fibres in the lumbosacral spinal cord of the cat, and the dendritic organization of the motor nuclei. Progr. Brain Res. 11, 120—152 (1964).

Staal, A.: Subcortical projections on the spinal grey matter of the cat. Thesis. Leiden: Koninklijke Drukkerijen Lankhout-Immig N.V.-S-Gravenhage 1961. 164 pp.

Szentágothai-Schimert, J.: Die Endigungsweise der absteigenden Rückenmarksbahnen. Z. Anat. Entwickl.-Gesch. 111, 322—330 (1941).

Torvik, A., and A. Brodal: The origin of reticulospinal fibers in the cat. An experimental study. Anat. Rec. 128, 113—138 (1957).

Tschermak, A.: Ueber den centralen Verlauf der aufsteigenden Hinterstrangbahnen und deren Beziehungen zu den Bahnen im Vorderseitenstrang. Arch. Anat. Physiol., Anat. Abth. 291—402 (1898).

Walberg, F., O. Pompeiano, L. E. Westrum, and E. Hauglie-Hanssen: Fastigioreticular fibers in the cat. An experimental study with silver methods. J. comp. Neurol. 119, 187—199 (1962).

Wall, P. D.: Cord cells responding to touch, damage and temperature of skin. J. Neurophysiol. 23, 197—210 (1960).

Woolsey, C. N.: Organization of somatic sensory and motor areas of the cerebral cortex. In: Biological and biochemical bases of behavior, p. 63—81. Eds. H. E. Harlow and C. N. Woolsey. Madison: The University of Wisconsin Press 1958.

Subject Index